Routledge R

Waste Location

First published in 1992, *Waste Location* seeks to widen and integrate the debate on the intrinsically spatial nature of waste disposal. The political and industrial significance of the new environmentalism of the 1980s came from the recognition of growing public pressure for environmental quality and product reliability. Attention was turned to waste as the product of consumption. As the political economy of waste was explored, new issues were raised: new technologies, recycling, pollution havens, waste minimization, location of landfill sites and incinerator facilities, and environmental crime, responsibility and planning. The 1990s sees the advocates of 'cradle to grave' responsibility still battling the promoters of market forces.

One of the major developments in the study of waste collection and disposal was the new forms of data collection and handling technology. The contributors consider both geotechnics and geographical information systems within this context. The focus on the geography of the UK is set within the broader framework of political economy and the international trade in pollution exports. The case studies presented range from bin analysis through a Bayesian perspective on risk to the global politics of international waste streams. Together, the contributors provide a comprehensive overview of the waste location debate in the early 1990s. Students of environment and climate change will find this book particularly enlightening.

Waste Location

Spatial aspects of waste management, hazards and disposal

Edited by Michael Clark, Denis Smith and Andrew Blowers

First published in 1992
by Routledge

This edition first published in 2021 by Routledge
2 Park Square, Milton Park, Abingdon, Oxon, OX14 4RN
and by Routledge
605 Third Avenue, New York, NY 10017

Routledge is an imprint of the Taylor & Francis Group, an informa business

A Library of Congress record exists under LCCN: 91009809

ISBN: 978-1-032-14532-7 (hbk)
ISBN: 978-1-003-23981-9 (ebk)
ISBN: 978-1-032-14531-0 (pbk)

Book DOI 10.4324/9781003239819

Waste location

Spatial aspects of waste management, hazards and disposal

Edited by
Michael Clark,
Denis Smith
and Andrew Blowers

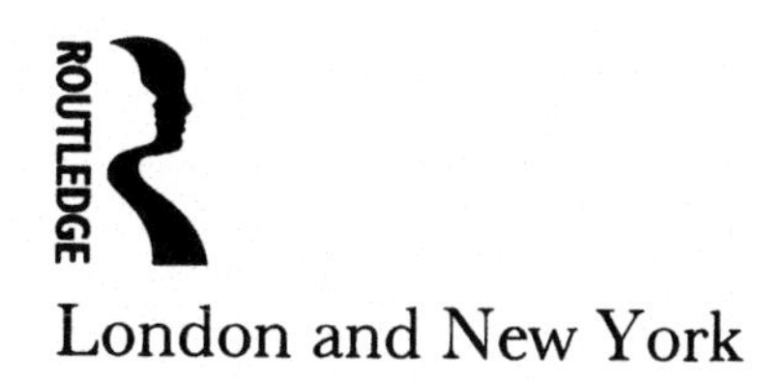

London and New York

First published 1992
by Routledge
11 New Fetter Lane, London EC4P 4EE

Simultaneously published in the USA and Canada
by Routledge
a division of Routledge, Chapman and Hall, Inc.
29 West 35th Street, New York, NY 10001

Typeset in Baskerville by Leaper & Gard Ltd, Bristol
Printed and bound in Great Britain by
Biddles Ltd, Guildford and King's Lynn

British Library Cataloguing in Publication Data
Waste location : spatial aspects of waste management,
hazards and disposal.
1. Waste materials. Disposal
I. Blowers, Andrew II. Clark, Michael *1949–* III. Smith,
Denis
363.728

ISBN 0–415–04824–9

Library of Congress Cataloging in Publication Data
Waste location : spatial aspects of waste management, hazards, and
disposal / edited by Michael Clark, Denis Smith, and Andrew Blowers.
p. cm. — (The Natural environment—problems and
management series)
Includes bibliographical references and index.
ISBN 0–415–04824–9
1. Refuse and refuse disposal. 2. Geographic information systems.
3. Refuse and refuse disposal—Great Britain. I. Clark, Michael,
1949– . II. Smith, Denis. III. Blowers, Andrew. IV. Series.
363.72′8—dc20 91–9809
CIP

Contents

Figures

Tables

Notes on contributors

K.G.M.M. Alberti, Department of Medicine, University of Newcastle upon Tyne, NE1 7RU.

Andrew Blowers, Professor of Social Sciences, Open University, Milton Keynes, MK7 6AA.

E.M. Bridges, Senior Lecturer in Geography, University College, Swansea, SA2 8PP.

Robert Brown, Research Assistant, Civic Amenity Waste Disposal Project, Luton College of Higher Education, Park Square, Luton, LU1 3JU.

Steve Carver, researcher, North-East Regional Research Laboratory, Centre for Urban and Regional Development Studies, University of Newcastle upon Tyne, NE1 7RU.

Michael Clark, Senior Lecturer in Geography, Centre for Environmental Management, Lancashire Polytechnic, Preston, PR1 2TQ.

P.C. Coggins, Senior Lecturer and Director, Civic Amenity Waste Disposal Project, Luton College of Higher Education, Park Square, Luton, LU1 3JU.

A.D. Cooper, Principal Lecturer, Luton College of Higher Education, Park Square, Luton, LU1 3JU.

Linda Crichton, Divisional Manager, Aspinwall and Company, Walford Manor, Baschurch, Shrewsbury, SY4 2HH.

Y. Crow, Department of Medicine, University of Newcastle upon Tyne, NE1 7RU.

Anthony C. Gatrell, Senior Lecturer, Department of Geography, Lancaster University, LA1 4YB.

P.T. Kivell, Senior Lecturer in Geography and Director of the Centre for Regional Information and Research, University of Keele, ST5 5BG.

Andrew A. Lovett, Lecturer, School of Environmental Sciences, University of East Anglia, Norwich, NR4 7TJ.

Stan Openshaw, Centre for Urban and Regional Development Studies, University of Newcastle upon Tyne, NE1 7RU.

Julian Parfitt, Senior Research Associate, Environmental Risk Assessment Unit, University of East Anglia, Norwich, NR4 7TJ.

Simon Raybould, Centre for Urban and Regional Development Studies, University of Newcastle upon Tyne, NE1 7RU.

Jonathon Renouf, television producer – chapter prepared while a postgraduate student at Durham University.

Trevor A. Sheldon, Department of Public Health Medicine, University of Leeds.

Denis Smith, Professor of Management, Liverpool Business School, Mount Pleasant Buildings, Liverpool, L3 5UZ.

Abbreviations

AIV	Advanced Interactive Video
BAND	Bedfordshire Against Nuclear Dumping
BATNEEC	Best Available Technology *Not Entailing Excessive Cost*
BGS	British Geological Survey
BNFL	British Nuclear Fuels
BOND	Britons Opposed to Nuclear Dumping
bpm	best practicable means
CAR	Conditional Autoregressive
CCMS	Committee on the Challenges for Modern Society
CCT	Compulsory Competitive Tendering
CEGB	Central Electricity Generating Board
CIPFA	Chartered Institute of Public Finance and Accounting
CPA	Control of Pollution Act
CPU	Central Processing Unit
DIY	Do It Yourself
DoE	Department of the Environment
DPWA	Deposit of Poisonous Wastes Act
DTI	Department of Trade and Industry
ECPR	European Consortium for Political Research
ED	Enumeration District
EHO	Environmental Health Officer
EPA	Environmental Protection Agency
ESRI	Environmental Systems Research Institute
fba	furnace-bottom ash
GAM	Geographical Analysis Machine

GIS	Geographic Information Systems
GLC	Greater London Council
HLW	High-Level Wastes
HMIP	Her Majesty's Inspectorate of Pollution
HWI	Hazardous Waste Inspectorate
IAEA	International Atomic Energy Association
IARC	International Agency for Research on Cancer
ICI	Imperial Chemical Industry
ILW	Intermediate-Level Wastes
IPC	Integrated Pollution Control
IWM	Institute of Wastes Management
JIS	Joint Information System
LAWDAC	Local Authority Waste Disposal Company
LLW	Low-Level Wastes
LULU	Locally Unwanted Land Uses
LWRA	London Waste Regulation Authority
MAUP	Modifiable Areal Unit Problem
MCE	Multi-Criteria Evaluation
MEL	Midland Environment Ltd
MoD	Ministry of Defence
NCB	National Coal Board
NGR	National Grid Reference
NII	Nuclear Installations Inspectorate
NIMBY	Not In My Back Yard
NIREX	Nuclear Industry Radioactive Waste Executive
NSCA	National Society for Clean Air
NWA	Northumberland Water Authority
OECD	Organization for Economic Co-operation and Development
OPCS	Office of Population Censuses and Surveys
PCB	Polychlorinated Biphenyls
pfa	pulverized fuel ash
PVC	Polyvinyl Chloride
RCEP	Royal Commission on Environmental Pollution
RDF	Refuse Derived Fuel
SHDC	Seaham Harbour Dock Company
SIAM	Society for Industrial and Applied Mathematics
SMR	Standardized Mortality Ratio

SSEB	South of Scotland Electricity Board
SVF	Single Variable File
TNC	Transnational Corporation
UKAEA	United Kingdom Atomic Energy Authority
USDA	United States Department of Agriculture
WDA	Waste Disposal Authority
WSL	Warren Spring Laboratory

Chapter 1

Paradise lost? Issues in the disposal of waste

Michael Clark and Denis Smith

> The true paradises are paradises we have lost
>
> (Marcel Proust, *Le Temps Retrouvé*)

> The world was all before them, where to choose
> Their place of rest, and Providence their guide:
> They hand in hand with wandering steps and slow
> Through Eden took their solitary way.
>
> (John Milton, *Paradise Lost*)

The 1990s are being hailed as the decade of the environment, as industry and government alike recognize the growing importance of public pressures for improvements in environmental quality and product reliability. The emergence of the new environmentalism during the 1980s had repercussions for all political parties in the UK. Every party has developed a green tinge to their policies in an attempt to capture the growing number of the electorate who wish to be seen as environmentally friendly. Within the media, numerous documentary programmes have extolled the virtues of a cleaner environment and the 'doom-watch' spectre has again risen to haunt both industrial and developing nations. The marketing and public-relations functions of industry have also risen to the challenge of making companies, along with their associated products, environmentally friendly in an attempt to appeal to the discerning green consumer. We cannot walk down the high street or watch television without being assailed by a range of products which espouse environmental improvements in addition to their primary functions.

Whilst the impact of the new environmentalism has been felt across the full range of industrial activities, its force has been felt most acutely within the production and disposal of chemicals. For the waste disposal industry the environmentally-conscious 1990s may prove to be a case of 'Paradise Lost'. The days are gone when both local publics and governments alike would tolerate waste disposal within their respective boundaries, especially if that waste is being imported from outside the region or locality. The body politic has become more educated as to the dangers inherent in such activities and has ensured that the disposal of waste is no longer a simple matter of finding a suitable hole in the ground. The days of compliant communities ended following the realization that the Love Canal débâcle in the USA was not an isolated aberration, but rather a spectre that hung over a number of communities which played host to waste treatment and disposal facilities. In the UK, the methane explosion at the Loscoe tip illustrated the dangers inherent in the disposal of supposedly 'innocuous' domestic waste – 'out of sight' was no longer 'out of mind'.

Whilst waste disposal companies still return handsome profits and display attractive earnings per share for the investor, it seems likely that such an euphoric state will be short lived as legal and political regimes become more hostile to the waste disposal function. The signs are already there that waste companies will find it difficult to escape intense local scrutiny and may, consequently, find it impossible to secure planning permission for new development. This has certainly been the case in the USA where the waste issue has held a longer tenure of notoriety than it has in Europe. The US disposal industry has witnessed a series of environmentally impacting events, suffered from scandals linking companies with organized crime and saw an upsurge in public concern for the environment; all of which conspired to give the industry an image of being 'the Frankenstein of the 1990s' (Smith 1990b).

During the late 1980s the issue of environmentalism was seen by some as being a potential driving force for those industries seeking to secure a market advantage over their competitors. As the 1990s develop it seems likely that the requirement on industry to be environmentally aware will become a reality through increased legislation and will require more than simply paying lip service to the 'green god(dess)'. As a consequence, environmentalism may well prove to be a major constraint on industrial production during the decade as public and political groups demand that corporate rhetoric is transformed into 'product reality'. Within the chemicals industry this requirement is already an emergent problem for industry, as major conflicts continue to arise over the production, storage and disposal of hazardous materials. In particular, the disposal of waste has emerged to become a major issue in the UK, mirroring the earlier development of the problem in the USA, as local public opposition becomes amassed against having disposal sites located 'in their backyard'.

How society deals with its waste is a good indication of political sophistication and priorities, and perhaps even a measure of social and planetary responsibility. Modern society's technical progress has been remarkable for the lack of attention, low status and limited resources given to the minimization, disposal or reclamation/re-use of 'waste' products, and to the analysis and control of their side effects or externalities. Political and economic structures have tended to minimize the value of materials which are discarded along with those environments which become degraded as a result of operations within the production–consumption–disposal cycle. They have also tended to give little, if any, value to those lives and interests harmed as a result of the process of economic development.

This dismal view of material progress preceding environmental and social responsibility (and to some extent being dependent upon its continued suppression) does at least allow for increasing environmental sensitivity. It is compatible with far less critical interpretations of the introduction of public health and pollution control measures. Although the adequacy of their effectiveness may be brought into question, these measures are most likely to be seen as an improvement on previous practice and priorities. In effect, a learning process is underway in which nation states begin to reconcile the competing moves towards increased economic prosperity and a better environment. Despite one's standpoint this has obvious implications for policy and is, to some extent, a reflection of *local* circumstances. Waste location – *the geography of rubbish* – is far more than an unfashionable and, until recently, largely ignored aspect of economic and transport geography and land-use planning. It has become a crucial part of the politics of the environment which has in turn led to a growing intolerance of irresponsible or dangerous behaviour. Indeed, such concerns have led to the criminalization of certain 'acts against the environment' and have helped to 'green' the public (although not necessarily the private) utterances of a remarkably wide range of organizations. It also threatens the continuance of a range of industrial activities that rely upon the local environment to accommodate or 'dilute' by-products that would otherwise require costly storage or treatment.

Recent UK experience within this 'hostile' political setting amounts to circumstantial evidence in support of a benign 'learning curve', within which there is a growing intolerance of 'cowboy economics' in determining material gain. This has been particularly evident in scientific and political discourse over the future of a number of common assets such as the oceans (polychlorinated biphenyls uptake in mammals, waste dumping), the atmosphere (destruction of the ozone layer, global warming) and land (waste disposal, pesticide use) within the context of competing economic activities. The degree of scientific uncertainty inherent within these debates has fuelled public concern and media speculation about the likely effects of continued environmental degradation. The resulting political pressure has forced industry to modify its strategic planning process in order to take account of these issues.

It is of little importance that this admonishment has taken place as a result of external pressure and popular concern. Nor does it matter that science has been caught up in processes of adversarial politics and interest group conflict, except where this situation has threatened to corrupt, limit or otherwise misuse independent professional rigour (see, for example, Smith 1990a). The lessons of the last few years mainly concern the costs of ignorance (scientific as well as political) because the failure to fully appreciate the harmful effects of degradation means that they continue unabated. Uncertain causal links and pathways suggest parallels with the idea of 'victimless crime', except here the victims are real; it is only the transgressor that remains anonymous. Such corporate anonymity exists as a function of the technical problems inherent in determining causality relationships. These uncertainties contrive to make it difficult to identify the perpetrator, as in the case of environmental cancers, because causality is often difficult if not impossible to prove beyond reasonable doubt.

This relationship between industrial development and environmental degradation is, of course, not a new phenomenon. Both the quality and length of life so many enjoy today would not have been possible without the unpaid sacrifice of generations of factory workers, miners and seamen, along with their families. Even today, the Third World's substandard living and working environments serve to go some way towards maintaining the developed world's much more secure life-styles. Such spatial inequality may be seen as an injustice, a system of unequal dependency which is manipulated by the powerful whose interests might be harmed by any improvement in the lot of the poor. Alternatively, it may be argued that any such improvements must be earned – that Third World people can have a clean and pleasant environment, but only once they have 'earned' it through the, necessarily dirty, process of economic development. Waste disposal politics is, therefore, a confrontation between those who argue that the process of development limits benefits to those who can *afford* a safe and pleasant environment, and those who argue that there should be an inalienable *right* to such benefits. The latter group maintain that those who threaten individual health or amenity as a result of their economic activity are involved in an indirect form of theft. The disposal of waste within the ecosystem of others can be seen therefore as a form of theft, but a theft which is both subtle and where causality is often difficult to prove.

While popular sentiment probably favours the idea of environmental rights, our institutions and economic conventions still reflect earlier values. 'Commons' such as the oceans, atmosphere and land are available for economic exploitation as long as no interests are harmed, and here the tendency is to value economic interests – fish stocks, game, tourism – above such intangibles as human health, aesthetic appreciation or ecological diversity. Whilst regulation has made great strides in controlling the worst excesses of 'local' air and water pollution, it has taken a fear of global catastrophe – acid rain, sea-level rises, ozone depletion – to stem some of the larger-scale discharges, and controversy still rages about the risks associated with radioactive and toxic emissions. Pollution pathways are often obscure, with interpretations of environmental quality ranging from those who argue that human life has never been so long or so rich, and that this has only been possible because of modern industrial products, to those who argue that we are living diminished and curtailed lives on a poisoned planet. Whatever the objective reality, popular sensitivity now includes widespread awareness of the dangers of proximity to certain types of hazardous material – be it radioactive, chemical or biological. Although industry and government are still criticized for sharing a cautious and environmentally damaging approach (for example BATNEEC: Best Available Technology *Not Entailing Excessive Cost*), gamekeepers and poachers have been distanced. Regulatory authorities are acquiring teeth and respect at a time when environmental law is likely to supersede more comfortable – and potentially corrupt – relationships.

It is tempting to suggest that the 'greening' of political culture in the late 1980s was a popular and widespread rejection of the incompetence, injustice and shortsighted stupidity of earlier years. Waste-related events, such as methane explosions, groundwater pollution and the internal trade in toxic waste, have contributed to widespread dissatisfaction with what may appear an *ad hoc* and badly-integrated approach to waste management and location. The revelation in the press during 1990 of a 'toxic time bomb', in the form of hundreds of inadequately-recorded and little-monitored landfill sites (actually the publicizing of surveys undertaken in 1973, and subsequently obscured rather than confidential), led to public concern of widespread soil contamination and expressed fear over the apparent association between various types of emission and the localized concentration of diseases. Such fears point to the need for urgent and independent research into the relationships between pollution and health, even though the findings could well create major political problems and will increase the cost and difficulty of waste disposal. Further drastic policy changes should follow damning criticism of land-reclamation projects in which the treatment of contaminated land has been influenced more by desired after-use than by the properties of the dangerous materials which have been buried. Imposing 'cradle to grave' responsibility should combine with more comprehensive and reliable monitoring of waste streams, especially across frontiers, to outlaw any routine dumping of toxic materials. Such responsibility should also serve to increase the stockpiling and disposal costs so that alternative, less damaging, materials and technologies are adopted.

While the overriding drift is towards increased cost, a 'new economics' perspective helps counter the impression that this is some burden of profligate environmentalism. The rising cost of disposal is in part a reflection of the working of market forces in allocating scarce landfill sites. It reflects the capital and labour costs required for the treatment, movement and containment of waste along with the extra need for effective monitoring and policing. The economic shift is also a powerful motivation for waste reduction at source: designing-out the 'throw away' element, and avoiding the creation of materials which are difficult or hazardous to dispose of. The cost of a shift towards source reduction would prove to be less than the burden which proper disposal entails. The waste disposal industry is currently suffering from a 'crisis of management' (see Smith 1990c) in which the respectable companies become tarnished by the inadequacies and malpractices of the 'cowboy' operators. Bad practice in the management of waste tips makes watercourses unusable, sterilizes adjacent land because of the danger of gas seepage and is often associated with lesser irritations: wind-blow material, vermin, odours and smoke. At worse, landfill sites contain lethal cocktails, have inadequate records and no proper monitoring after completion. Bad practice in incinerator management threatens to scatter toxic material over neighbouring areas and create hazardous ash, whether or not the furnace was burning toxic materials. The very worst waste management practices kill, through avoidable disasters, workplace accidents and disease. The pathways for many of these bad effects are so complex, obscure or open to question that it is perhaps best to be cautious.

The publication, in October 1990, of the Government's White Paper illustrates the dynamic uncertainty which surrounds waste issues. The emphasis presently placed upon waste minimization, recycling and the adoption of clean technologies is now set firmly within the political context of global responsibility. BATNEEC is interpreted in a way which suggests that regulatory bodies will pay more attention to environmental outcomes than to commercial considerations, with the regulators being encouraged to improve standards in the light of technological advances (Department of the Environment 1990). There are also indications that the UK Government now favours international self-sufficiency and the minimization of hazardous waste transfers. However, the stated policy objective that the UK should not import wastes for direct landfill still leaves open the possibility that the more highly toxic waste streams may still be imported for incineration and treatment. A new-found official enthusiasm about the effects of a strict regulatory regime on waste generation is paralleled by a more general reliance on market forces and by renewed optimism over the regulatory agencies' abilities to cope. The 'polluter pays principle' and Her Majesty's Inspectorate of Pollution's (HMIP's) associated new powers to levy charges on polluters are optimistically expected to be adequate in overcoming the agency's well-publicized shortage of inspectors. Similarly, the White Paper's statement that there is 'no shortage of suitable sites in most parts of the country' (ibid.: 193) ignores the planning dynamic, the 'Not In My Back Yard' (NIMBY) syndrome and the problem of local accountability. The political dynamics of the problem are sufficiently complex to prevent a speedy acquiescence by local politicians and environmental pressure groups – political and legal rhetoric about environmental improvements will need to be quickly translated into measurable improvements in order to satisfy the increasingly restless electorate.

WASTE DISPOSAL AND THE SPATIAL DYNAMIC

Our aims in this book are to attempt to inform and widen debate about a range of waste-related issues, to help the process by which waste disposal is being integrated into the UK land-planning system and to promote and enhance the contribution which geographers have made in the field of waste disposal studies. Here the book seeks a wider audience; partly because geography can be a useful mediator between science and politics by increasing the accessibility of work within the more highly-specialized and technical disciplines and then helping relate this to policy. Hopefully this adds value as well as popularizing. It incorporates the discipline's specialisms of spatial analysis and a rigorous approach to the spatial dimension in studies which cross disciplinary and professional boundaries, including both physical and social sciences. Although they represent a fair sample of current geographical interest in waste issues, each of the chapters is an independent piece of work and authors do not necessarily agree.

The range of geographical studies is reflected in the book's structure. This runs from the concrete and practical contribution of physical geography, via information systems and the controversy about patterns of disease, to political issues at a variety of scales. *Geotechnics* is represented by chapters on the contribution of waste disposal to land reclamation and on the availability of landfill cover materials in South Wales. Kivell explores the relationship between landfill and derelict land. What should be a convenient way of turning problems into solutions turns out to be a complex and difficult set of circumstances made worse by a lack of co-ordination. Bridges, in turn, deals with the specific issue of locating suitable landfill cover. Even an area as geographically diverse and extensively mined, quarried and excavated as South Wales faces sufficient difficulty for some co-ordination of civil engineering programmes and landfill operations to be recommended.

The importance of *Geographic Information Systems* (GIS) is reflected in its coverage over four chapters. Parfitt discusses the use of special Waste Consignment Notes for the Greater London area. Such data illustrate the complexity and variability of waste records and they also show how difficult it is to obtain comprehensive and reliable information. Abolition of the Greater London Council (GLC) added to these difficulties, paradoxically at a time when the enormous task of coping with London's toxic waste was becoming more widely appreciated. Effective policies require better data and far more co-ordination than has been achieved to date.

Crichton outlines the advantages of a geographical information systems approach for two specific applications. Aspinwal's Waste Information Database provides a computerized database which permits users to identify licensed sites in terms of a wide variety of classifications, for both operational and closed sites. Their refuse collection planning and management system should help the operational efficiency of domestic waste collection by applying, among other findings, the relationship between bin content and households' socio-economic classification.

What we throw away can be predicted, and the gathering and use of such information is also central to the chapter by Coggins, Cooper and Brown from the Civic Amenity Waste Disposal Project at Luton College of Higher Education. Their research has given particular attention to public use of household waste disposal facilities: what is taken to these (oddly-named) civic amenity sites, how far, how often and by whom? Patterns of disposal are shown to be strongly affected by the introduction of larger capacity wheeled bins. This has implications for the viable operation of civic amenity sites, and the prospect of charges being introduced for some categories of domestic waste raises further, and more political, issues.

Carver and Openshaw then use the contentious example of nuclear waste site location to explore the potential of a GIS approach to site identification and evaluation. It is argued that NIREX (Nuclear Industry Radioactive Waste Executive) has not made full use of the scope for comprehensive site-searching, criteria modification and 'what if' modelling that a GIS approach can offer. If NIREX wish greater understanding of the site-search process, and to overcome the powerful NIMBY hostility to location anywhere in the UK, they should make more use of GIS.

Discussions of *Spatial Association* address the epidemiological uncertainty which has become an important reason for local hostility to toxic and radioactive waste disposal facilities. Raybould, Crow and Alberti explore the relationship between heavy metals in soils and the distribution of diabetes. Data for the Tyne and Wear area point to spatial associations, though research is still at an early stage and its main implications are for the direction of further investigation. Gatrell and Lovett also face severe data and methodological constraints in their study of the health implications of incineration. Circumstantial and anecdotal evidence, political controversy and the threat of legal proceedings against any unsubstantiated accusation all hinder academic enquiry and debate. The example of the possible relationship between a Lancashire incinerator and the distribution of cancer patients illustrates the alarmist nature of media reaction, the great inadequacy of UK population and health data and the danger of 'spot the ball' cluster identification around points such as incinerators. Even so, the example does warrant further study. Some evidence points to an association between the incinerator's location and the incidence of cancer of the larynx, but it will take many years and far better information to establish whether or not a causal link exists.

Sheldon and Smith consider the methodological problems that beset research in the field of epidemiology and offer an alternative research scenario, through the use of Bayes's Theorem, for subsequent studies in this field. They also consider the difficult relationships that exist between scientific knowledge and the political interpretation of such information. It is at this interface between science and society that geographers have the potential and the opportunity to act as mediators between opposing viewpoints.

The section concerning *Political Economy* focuses on decisions and political processes at two contrasting scales. Renouf's study of colliery spoil-dumping on Durham beaches shows local objection to loss of amenity and fishing grounds combined with growing environmental objections, but it is countered by coal industry cost minimization in the face of closure threats. Smith and Blowers investigate the international politics of the hazardous waste trade where it appears that, despite attempts at regulation, these waste streams are inadequately recorded and to a large extent controlled by the profitability of low cost destinations. This favours Third World disposal and various forms of bad practice. However, media attention to 'leper ships' and toxic waste eco-crime may help create an international political climate in which safety rather than profit is paramount.

Blowers's final chapter takes a wider perspective on the politics of waste. To what extent will disposal difficulties limit the *generation* of waste product? Can industry reduce the amount of material that is difficult or hazardous to dispose of? Might increasing political sensitivity and more rigorous regulation require an increase in the amount that needs special treatment of disposal facilities? What about the toxic legacy of materials already discarded? Do old landfill sites represent a time bomb which can only be defused at great cost, and with the effect of further reducing the availability of sites for future disposal use? In the early 1990s the UK is experiencing some of the hostility to dangerous and locally unwelcome waste disposal that has been important in North American environmental politics for at least two decades. It is also under European Community pressure to adopt strict emission and discharge standards rather than the long-established, negotiated and professionally-assessed approach to individual licences.

Integrated pollution control and the incorporation of statutory waste disposal plans, within the structure planning framework of land-use management, promise a shift away from public health nuisance control and self-regulation. However, our environmental policemen are notoriously short staffed and lack the status of, say, New Jersey's eco-crime busting 'Green Police'. It is a sick paradox that Greenpeace's attempts to expose dangerous and illegal discharges by UK companies have been met by the full force of the law: imprisonment, heavy fines and injunctions. The great difficulty of proving that emissions cause harm, the lack of legal protection for amenity or peace of mind, and the far greater financial resources of large corporations all mitigate against faith in legal safeguards. However, even here there are signs of progress. The selling-off of nationalized industries has been accompanied by various measures to separate 'poachers from gamekeepers'. It is no longer acceptable for the same organization to combine delivery and regulatory functions. Freedom of information conventions, though weakly developed in the UK, are beginning to reflect popular objection to the overriding status of commercial interests and confidentiality – especially when this is used to hide wrongdoing and criminal behaviour.

REFERENCES

Department of the Environment (1990) *This Common Inheritance*, London: HMSO.

Smith, D. (1990a) 'Corporate power and the politics of uncertainty: conflicts surrounding major hazard plants at Canvey Island', *Industrial Crisis Quarterly* 4 (1), 1–26.

Smith, D. (1990b) 'The crystal ball syndrome – the strategic importance of green issues', Paper presented at the conference 'Greening the curriculum: implications of environmentalism for business education', organized on behalf of the Business Education Teachers' Association, Leicester, 26–27 April 1990.

Smith, D. (1990c) 'Beyond contingency planning: towards a model of crisis management', *Industrial Crisis Quarterly* 4 (4), 263–75.

Chapter 2

Land reclamation through waste disposal

P.T. Kivell

INTRODUCTION

At first glance there may appear to be a neat symmetry between waste disposal and land reclamation. Here is the possibility for two problems to be put together in order that each may provide a solution for the other. Traditionally over three-quarters of Britain's refuse has been dumped in landfill sites and most waste disposal authorities would argue that this nearly always plays a land reclamation role. Indeed, some would claim that restoration of land is one of the waste industry's proudest achievements (Reeds 1987). However, closer examination reveals that the problems are neither as simple, nor as self-cancelling as this might imply. Not all waste is suitable for landfill purposes, not all derelict land is physically suitable for waste disposal and the quantities and locations of the two rarely match.

This chapter is mainly concerned with the characteristics of the waste and the land, and the ways in which they can be matched for the restoration of derelict sites. First, however, it is necessary to review briefly the institutional framework, and particularly the changes in the past fifteen years which have encouraged land restoration.

Since 1974 in England both the planning and the licensing of waste disposal have been broadly the responsibility of the counties, although statutory authorities operate in some of the metropolitan areas and the districts have prime responsibility in Wales. This has resulted in improved co-ordination. For example, in West Yorkshire before 1974 there were fifty-four local authorities responsible for their own waste disposal and most of them were too small to finance capital investment or technical advances. Since 1974 West Yorkshire has progressed further than almost any other authority by setting up a combined Waste Management and Land Reclamation Unit (Sims 1984). The other main institutional changes nationally concern the tightening of regulations by the Control of Pollution Act 1974, which imposed stricter site licensing conditions, and more recently the work of Her Majesty's Inspectorate of Pollution which will be referred to later.

The waste

Sophisticated modern societies produce many forms of waste. In general these may be classified according to their source, for example, domestic, manufacturing industry, power generation, mineral extraction, or according to the type, for example, containing hazardous matter, biodegradable matter or inert material. A number of these wastes have very specific characteristics which exclude their use for land reclamation, so the emphasis in the

Table 2.1 Contaminants and the problems posed for land reclamation

Type of contaminant	*Likely occurrence*	*Hazards*
Toxic and other metals (lead, cadmium, mercury, arsenic, copper etc.)	Metal mines, engineering works, electroplating works, industrial waste	Harmful to humans, animals and plants
Combustibles (coal, coke-dust)	Gas works, power stations, railway land	Underground fires
Flammable gas (e.g. methane)	Landfill sites, domestic waste	Explosions within or beneath buildings
Aggressive substances (sulphates, chlorides, acids)	Landfill sites, industrial-waste-made ground, slag heaps	Attacks building materials
Oils, tars, phenols	Chemical works, tar distilleries	Contamination of water
Asbestos	Industrial buildings and waste sites	Harmful if inhaled

Source: Based on Table 1, Interdepartmental Committee on Redevelopment of Contaminated Land (1987).

present chapter will be upon the broad range of domestic and industrial materials. Table 2.1 gives an indication of the main problematical substances.

Waste disposal statistics reveal that a total of approximately 25.1 million tons of domestic, industrial and commercial waste was dealt with by the waste disposal authorities in England in 1986–7 (Hunt 1988). This figure is necessarily an approximation, for less than half of the waste is actually weighed, and there are additional amounts, especially of industrial waste, which are disposed of, or recycled, privately. What is clear is that the upward trend in waste, evident for many years, is currently running at about 5.6 per cent per annum. Other estimates for the generation of purely domestic waste suggest that we are now each responsible for 350 kg/year (DoE 1986a).

The nature of this waste has changed appreciably in recent decades and this has major implications for landfill. In particular, since the 1950s the proportion of coal ash has fallen markedly and that of paper and other biodegradable materials has increased. More recently still the addition of 'modern rubbish' in the form of plastics and polystyrenes has brought a new range of problems. About 50 per cent of domestic rubbish is thought to be combustible (Bradshaw and Chadwick 1980) but the mixture is very variable and costs are relatively high. Consequently there has been a reappraisal of incineration and the closure of a number of large plants, for example, at Whetstone in Leicestershire and at Bidston Moss on Merseyside.

Methods of disposing of this waste vary according to local circumstances but in general, sanitary landfill remains the most cost effective method as illustrated by figures for Merseyside given in Table 2.2. In the 1970s, 78 per cent of domestic refuse was landfilled (Mundy and Gaskarth 1979) and this remains broadly true today.

Despite the rapid growth of environmental concern, and the tightening of regulations regarding waste disposal, there is disturbing evidence that many landfill sites are not being efficiently operated. A recent report by HM Inspectorate of Pollution (1989) made the following observations about conditions in some of the Metropolitan Counties: Tyne and Wear – 'licence conditions not being enforced with sufficient vigour or consistency by either districts or the central team' (p. 10); 'the lack of regulation of public sector landfill was illustrated in the examples of poor standards observed. Of particular concern was the lack of sub-surface monitoring for landfill gas where it was clearly evolving' (p. 11); South

Table 2.2 Comparative costs of waste disposal on Merseyside

Type of waste disposal	*Cost (£)*
Local landfill	4–7 tonne
Transfer and landfill	9–14
Pulverize and landfill	11–17
Resource recovery	12–17
Landfill via rail	16–20
Incineration	20–30

Source: Shimwell (1987).

Yorkshire – 'the effectiveness and enforcement of licences is questionable' (p. 19); West Midlands – 'a substantial percentage of these [private sector] sites fail to meet the standards set by the Department ... particularly in terms of the control of gas and leachate' (p. 23); Greater Manchester – 'arrangements in the Greater Manchester area are good' but even here 'the supervision of a number of landfills ... was inadequate' (p. 32).

Such findings reveal present-day shortcomings and point to likely difficulties in the future when such sites need to be reclaimed. The problem is therefore a twofold one comprising latent technical defects and a threat to public confidence in the landfill/land restoration process.

THE LAND

The definition of derelict land introduces a number of complications into the argument. Currently dereliction is defined as much in terms of planning and administrative criteria (that is, the likelihood of attracting Derelict Land Grant) as it is in terms of physical criteria or the degree of abandonment. In addition, sites on which mineral or domestic waste tipping is currently, or sporadically, taking place are excluded from dereliction statistics if they have planning permission and/or eventual restoration conditions attached. Department of the Environment figures show that in 1974, 11,400 ha of land were being used in England for non-mineral tipping, 75 per cent of which was subject to restoration conditions (DoE 1975).

Derelict land originates in a number of different ways and not all of it is suitable for tipping. Most obviously sites which offer a large void have the greatest potential for tipping, thus disused quarries, natural depressions (possibly subject to flooding), backfilling of mineral sites and the use of abandoned canals, docks and railway cuttings may all be considered. Having decided in principle that a void is suitable for filling, detailed engineering and geotechnical surveys may reveal constraints. Dry quarries offer a wide range of possibilities, but water poses many difficulties. For example, a clay quarry with adequate sealing and controlled drainage may be suitable for mildly toxic wastes, but if these conditions are not met, or break down, the site may become a source of pollution, producing toxic leachates.

In addition to artificially-created voids, there exist a number of marshes, mosses and other semi-natural low-lying areas with potential for tipping. In the past they have been readily used because their agricultural value has been virtually zero and they have been unsuitable for building or other development. Increasingly, however, their ecological and environmental values are being realized and this limits their use for waste disposal.

Finally, it is not always necessary to have a void for waste disposal. Generally waste disposal authorities in Britain do not tip above the surrounding ground level, except by a very small margin to allow for settlement. Occasionally, however, the formation of a mound or ridge, as part of a landscaping scheme, may incorporate quantities of waste material. Such was the case during the creation of the large central ridge at the Stoke-on-Trent Garden Festival Site in 1986. Similarly, in Bochum, West Germany, refuse has been dumped in a 'mountain', 120 m high, which provides a number of leisure facilities including a dry ski slope (Porter 1982).

MATCHING THE WASTE AND THE LAND

Two essential considerations guide the way in which waste may be used for land reclamation:

1 The distance between the source of the waste and the tipping site.
2 The compatibility of the waste and the disposal site.

Distance is crucial in the economics of waste disposal. Although specialized wastes of low weight may be transported half-way around the world for treatment or disposal, the overwhelming bulk of domestic and industrial waste is disposed of very close to its origins.

This immediately introduces a regional consideration into the discussion for there is not a particularly good match, even in crude terms, between the areas where waste arises and areas where there is derelict land which lends itself to reclamation by landfill. In a few areas there is an abundance of derelict voids suitable for filling. For example, Leicestershire with many opencast mining areas and a massive (4 mill, m^3) granite quarry at Enderby is even thinking of importing waste (Porter 1983). Elsewhere there may be a local balance, as around Leeds, where the rate of mineral extraction almost exactly matches the creation of industrial and domestic waste (Crosby and Renold 1974). A special problem exists, however, for London, which has been driven to use the Essex Marshes, clay pits in Bedfordshire and gravel pits at Didcot and elsewhere at distances of up to 80 km. Even some of these schemes have foundered on the grounds of costs for transporting the waste. More generally, even in the congested South-East it has been suggested by Philpott (1982) that there are sufficient landfill sites for thirty years. As land prices increase, however, competition for the holes also increases. For this reason some dry quarries, for example, may be more valuable for industrial building than for waste tipping.

The compatibility between the kind of waste material and the physical nature of the site is the second major consideration. As far as the waste is concerned the main limiting factors are potentially toxic and hazardous materials and the increasingly organic nature of waste. The first problem is an obvious, but very difficult one. Hazardous materials in domestic and industrial wastes are unpredictable but include a variety of chemicals, asbestos, glass and sundry pathogens. These pose both short-term dangers during tipping and longer-term threats to the local water regime and land quality. Organic materials present a variety of problems when used for landfill, but there are two general issues.

First, there is the problem of decomposition which leads to odour and attracts vermin, but in the longer term may also produce heat, moisture and potentially lethal gases, notably methane. Attention was drawn to this problem in a dramatic way by a methane explosion which destroyed a bungalow at Loscoe in Derbyshire in 1986. Such problems prompted HM Inspectorate of Pollution to publish a Waste Management Paper, no. 27. This, and associated circulars revealed that there are up to 1,400 landfill sites with potential gas problems and half of them are within 250 m of housing or industry. Methane may, of course, be vented and flared off, or used for small-scale power generation or to fire bricks, but its presence certainly limits building on, or close to, landfill sites. Second, there is the problem of shrinkage and consolidation, which can be up to 50 per cent with some kinds of organic refuse. This may necessitate elaborate compaction as outlined below.

Various forms of treatment may be applied to make waste more suitable for land reclamation. Initially the waste needs to be carefully classified and dumped according to the nature of the site. Some initial sorting and recycling may be undertaken in which toxic wastes are removed, metals are recovered and other materials are rendered less volatile. Toxic wastes may be sealed and loose, high bulk materials may be compacted or baled at a processing plant; but all of this adds to the costs. Hunt (1988) suggested that compacting crude waste costs between £3.45 and £4.18 per ton, and shredding is even more expensive at £5.77–£7.50. These prices were compared with averages between £1.16 and £2.60 for landfill disposal, but elsewhere considerably higher figures obtain for all of these categories. In the provinces, 72 per cent of waste is disposed of without treatment, but in London and some metropolitan areas, a shortage of landfill sites means that three times more waste is treated as is disposed of without treatment. This has obvious implications in terms of higher costs.

Turning to the land side of the equation, it is clear that here too some initial classification and subsequent treatment is necessary. Most importantly sites need to be classified according to the risks to public health and amenity and the dangers of water pollution. Since 1974 tipping has been governed by the Control of Pollution Act. Broadly, this requires that tipping should take place in compartments, preferably on a puddled clay floor, in layers no more than 2.5 m thick. Each layer should be compacted and covered with subsoil and a final layer of soil 1 m thick should cover the site (Bridges 1987). A degree of compaction may be achieved by the passage of contractor's vehicles but a number of specific techniques are available for increasing the density of the filled material. These include the use of vibratory rollers and dynamic consolidation by repeatedly dropping a heavy weight from a great height (Charles 1979). A fairly rapid compression may also be achieved by temporarily preloading the site with a surcharge of fill (Charles *et al.*, 1986).

The net value of waste disposal sites for land reclamation depends largely upon the balance between location and after-use capability (see Figure 2.1). This latter factor depends in turn upon the kind of

Value for waste disposal	LOW value for land reclamation	HIGH value for land reclamation
HIGH	Site able to take any kind of waste, close to origin. Poor afteruse potential for building	Site able to take many kinds of waste, within urban area
LOW	Site with restricted capabilities, able to take limited types of waste and in poor location for reclamation	Site able to take only inert waste Suitable for building development

Figure 2.1 Relative value of sites for waste disposal and land reclamation

Source: After Crosby and Renold (1974).

materials which a site can accommodate. Unfortunately sites which have a high value for waste disposal, that is, those which can accept almost any kind of waste, usually have a low capability for reclamation because of the limited range of after-uses which they will support. One of the commonest situations is for a site to have a combination of low waste disposal potential and high reclamation value. These are sites which can take inert waste only (such as builders' rubble or solid pottery waste), on which it is possible to build, but the availability of such waste is limited. In all of these cases planning guidance needs to consider a wide range of issues such as access, the relationship with adjoining areas, landscape values, wildlife, screening, drainage and the need for various after-uses.

Most landfill sites have serious limitations for re-use. The typical site is effectively a complex reactor containing physical, chemical and biological processes which may remain active for many years. This is especially true of the uncertain conditions created by the domestic and municipal tipping which took place before the mid-1970s. Even today the increasing amounts of domestic and industrial wastes are rarely placed and compacted as a controlled engineering operation. Consequently, the main potential for re-use lies with extensive, low ground-loading activities, such as agriculture, leisure and recreation, including golf courses.

Some building development is, of course, possible, but it needs to be undertaken with the utmost care. Consideration must be given to the nature, depth, age, and compaction of the fill and whether construction will damage impervious caps on contained fills. The British Standards Code of Practice for Foundations (CP 2004) recommends that 'All made ground should be treated as suspect, because of the likelihood of extreme variability' (Smith 1979: 363). The Department of the Environment echoes this advice by suggesting that 'building on restored landfill is generally not recommended at least for a considerable period after its completion, although following assessment, building on shallow old landfills may be possible' (DoE 1986b: 75).

The main problems as far as buildings are concerned involve biodegradation and the generation of gases, chemical hazards, the low load-bearing capabilities of landfill and the unpredictable compaction/settlement effects, especially on the margins between landfill and the original ground. Building regulations give only partial guidance on landfill, but a recent consultant's report (ECOTEC 1985) suggests that fifteen to twenty years should elapse before certain landfill sites are developed. There is presently a danger that the pressure to find building sites in some areas may be encouraging builders to take on filled sites without a full investigation. Recently at West Leigh, near Manchester, a development of fifty houses had to be halted when unforeseen methane gas problems occurred and in Exeter, Lovell and Norrish (1986) reported that a housing scheme had to be restricted for the same reason. Many of the smaller firms do not have the resources or expertise to deal effectively with the problems which may be encountered.

Chemical hazards are an unpredictable and potentially damaging feature of some landfill sites. In particular they may attack subsurface metal and concrete foundations, cause changes of volume in the waste material and lead to health hazards. Of course, a number of engineering techniques are available for stabilizing land, strengthening foundations and neutralizing chemical activity, but these remedial measures are not always effective and may even lead to further problems, for example, piled foundations may penetrate a sealing layer and provide channels for leachates.

Where building is to take place, it is desirable and most cost effective to consider its nature and extent at the outset, and to incorporate the necessary provisions in the overall landfilling operation. Many of the engineering considerations have been outlined by

Crawford and Smith (1985), and the DoE (1986b) point to the care that must be taken over foundations and safeguarding services. They estimate that the costs of restoring landfill sites for building range from £15,000–£30,000 per hectare (at 1984 prices) and that additional costs will be incurred for aftercare such as gas venting, the treatment of settlement and leachates and routine maintenance.

SUCCCESSFUL LAND RECLAMATION THROUGH WASTE DISPOSAL

Despite these manifold problems there are plentiful examples of successful land reclamation through waste disposal, a few of which date back to the nineteenth century. For example, the corporation of the City of Manchester began buying land at Carrington Moss and Chat Moss in the 1880s as a site for dumping night soil and other refuse. Reclamation of the mosses was not the original purpose, but eventually over 1,400 ha of valuable agricultural land was reclaimed in this way, creating a useful income for the city (Phillips 1980). More recently sewage sludge has been used, not as landfill but as a treatment agent to encourage revegetation for quarry reclamation schemes in Somerset and Derbyshire (Bradshaw 1982) and in strip-mining areas of the USA (Edgerton 1978).

Subject to the limitations noted above, domestic refuse has been widely used as a landfilling material for coastal or other wet sites, although in most cases elaborate structures are also necessary. Liverpool City Corporation reclaimed a large area of the frontage of the River Mersey in the 1920s by dumping municipal refuse behind a concrete wall to create the pleasant recreational area of Otterspool Promenade. Portsmouth, a city which has long suffered from a shortage of development land, reclaimed 164 ha of its shallow harbour area in the 1970s by controlled dumping of refuse behind artificially-created chalk embankments at Paulsgrove Lake. Similarly industrial and commercial development on the foreshore of Belfast Lough, where tipping has taken place since the 1950s, represents one of the largest schemes in the UK.

Even more ambitious schemes have been taking place over the past twenty years in an area known as the Hackensack Meadows in New Jersey. Here, under the guidance of a special Development Commission, an untidy stretch of 8,000 ha of marsh, tidal basin and landfill is being reclaimed through the dumping of 45,000 tons of garbage per week to create a recreational, commercial, industrial and residential facility on the outskirts of New York City (Grant 1976).

In Britain there is nothing on this scale but many urban authorities have successfully reclaimed land filled with refuse for agricultural, amenity or even building purposes. Amongst the best known of these are the attempts to dispose of London's waste in a constructive manner. In Essex, Corey Waste Management operates two large landfill sites, one at East Tilbury for industrial and commercial refuse, and the other at Mucking for domestic waste. A first phase of restoration for grazing land has already been completed (Fennell 1986). To the north and west of the capital other large landfill/land reclamation schemes are to be found. Since 1970 the London Brick Landfill Company has been bringing together the need for waste collection, disposal, and land restoration in the Bedfordshire brickfields. At Stewartby there is a twenty-year scheme to restore 80 ha to farmland using domestic refuse from London. Similarly, at Sutton Courtney in Oxfordshire gravel pits are being reclaimed with a mixture of London's waste and pulverized fuel ash from Thames Valley power stations.

Smaller, more specialized schemes have been undertaken elsewhere, notably in the New Towns of Corby, Telford and Warrington, and mineral wastes have also been used to rehabilitate land in a number of localities, notably Nottinghamshire (Macpherson 1983; Payne 1984). Perhaps the most comprehensive and best co-ordinated programme is that in West Yorkshire, where derelict land and waste disposal are tackled together in an overall recycling policy which was responsible for the restoration to beneficial use of 150 ha of land on forty-seven separate sites in the decade 1974–84.

CONCLUSION

A number of individually successful schemes can thus be cited, and yet the relationship between land reclamation and waste disposal remains mostly an *ad hoc* arrangement. We are still a long way from the recommendations of the 'Countryside in 1970 Committee' that 'All refuse tipping should be based on a planned scheme of land reclamation'. In addition to the operational, engineering and locational problems noted above, there remain problems of administrative co-ordination. Outside of the main urban areas, waste disposal is mainly a county function in England, and like most other land-using activities, it requires planning provision. Where land reclamation is a possibility, careful consultation is required between reclamation teams, land-use sections and those concerned with mineral extraction within the local authority, but also with adjacent local authorities, mineral operators, major waste producers, disposal contractors and regional water authorities. Such is the case with the large-scale Judkins landfill site operated by Amey Roadstone Corporation and Warwickshire County Council in a complex of quarries near Nuneaton. At this site 700 tonnes of domestic and quarry waste are being dumped each day with the eventual aim of creating a Country Park. Ideally, as here, the restoration techniques and the requirements of the projected after-use should be built into the programme from the outset. In the case of large schemes the process of tipping and land restoration may be phased over more than twenty years, so a degree of flexibility in the plans is also desirable.

The economics of waste disposal are still very much in favour of landfill, but its potential for land reclamation is limited by a number of factors. First, government land reclamation policy and grants favour building as the preferred after-use, but this is commonly not advisable on filled land. Second, many major cities are fast running out of landfill sites, especially those which might be useful for planned redevelopment. Third, the kind of material which is available in waste disposal operations increasingly rules out a number of after-uses on landfill sites unless it is first treated and then carefully maintained for a lengthy period. Fourth, the nature of the landfill process itself, especially before 1974, has created a legacy of problems, including subsidence, gas and leachates which render many sites unsuitable for intensive development. Almost all of the engineering problems of building in such circumstances can be overcome, but only at considerable cost. In the foreseeable future the findings of the Interdepartmental Committee on Redevelopment of Landfill Sites (1988) are likely to guide general policy. Principal among these are the suggestion that soft uses, for agriculture and recreation, are probably preferable, and the least expensive means of returning land to beneficial use.

REFERENCES

Bradshaw, A. (1982) 'The landscape reborn', *New Scientist*, 30 September, 901–4.

Bradshaw, A.D. and Chadwick, M.J. (1980) *The Restoration of Land*, Oxford: Blackwell.

Bridges, E.M. (1987) *Surveying Derelict Land*, Oxford: Clarendon Press.

Charles, J.A. (1979) 'Field observations of a trial of dynamic consolidation on an old refuse tip in the East End of London', *The Engineering Behaviour of Industrial and Urban Fill*, Proceedings of Midland Geotechnical Society Symposium, Birmingham, pp. E1–E14.

Charles, J.A., Burford, D.D. and Watts, K.S. (1986) 'Preloading uncompacted fills', Building Research Establishment Information Paper IP 16/86, Watford.

Countryside in 1970 Committee (1969) 'Refuse Disposal', Report of the *Ad Hoc* Committee on Refuse Disposal, *The Countryside in 1970*, London.

Crawford, J.F. and Smith, P.G. (1985) *Landfill Technology*, London: Butterworths.

Crosby, J.G. and Renold, J. (1974) 'Where to put solid wastes', Local Government Operational Research Unit, Royal Institute of Public Administration, Report no. 168.

Department of the Environment (1975) *Survey of Derelict and Despoiled Land in England, 1974*, Derelict Land and Waste (Refuse) Tipping Section, London: HMSO.

Department of the Environment (1986a) *Transforming our Waste Land: The Way Forward*, London: HMSO.

Department of the Environment (1986b) 'Landfilling wastes', Waste Management Paper no. 26, London: HMSO.

ECOTEC (1985) *The Re-use of Landfill Sites*, ECOTEC Research and Consulting Ltd, Birmingham.

Edgerton, B.R. (1978) 'Revegetating bituminous strip-mine spoils with municipal wastewater', *Compost Science* 16, 20–25.

Fennell, M.J. (1986) 'Domestic waste disposal', *Public Health Engineer*, 14 (2), 33–5.

Grant, R.C. (1976) 'Turning it around in the Hackensack Meadows', *Urban Land* 35 (10), 15–22.

Her Majesty's Inspectorate of Pollution (1989) 'Waste disposal regulation and operations in the former Metropolitan counties: a review', Occasional Paper, London: DoE.

Hunt, J. (1988) 'Waste disposal statistics', *Association of County Councils Gazette* 81 (1), 9–15.

Interdepartmental Committee on Redevelopment of Contaminated Land (1987) 'Guidance on assessment and redevelopment of contaminated land, 59/83', 2nd edn, London.

Interdepartmental Committee on Redevelopment of Land Fill Sites (1988) 'Guidance note 17/78' 7th edn, London.

Lovell, A.M. and Norrish, J.H. (1986) 'Redevelopment of a former landfill site', *Environmental Health* 94 (12), 323–7.

Macpherson, T. (1983) 'Rehabilitated land for landfilling operations', *The Planner* 69 (4), 130–1.

Mundy, M.J. and Gaskarth, J.W. (1979) 'Practical considerations of waste landfill operations', *The Engineering Behaviour of Industrial and Urban Fill*, Proceedings of Midland Geotechnical Society Symposium, Birmingham, pp. A9–A24.

Payne, V. (1984) 'How counties cope with mineral waste disposal', *County Council Gazette* 77 (3), 84–5.

Phillips, A.D.M. (1980) 'Mossland reclamation and refuse disposal in the Manchester area in the nineteenth century', *Industrial Archaeology Review* 4 (3), 227–33.

Philpott, M.J. (1982) 'Trends in waste disposal planning', in *Re-use of Solid Waste*, Institute of Civil Engineers Conference, London: Thomas Telford Ltd, pp. 97–104.

Porter, C. (1982) 'Germany builds leisure mountains from waste', *Surveyor* 160 (4717), 8–10.

Porter, C. (1983) 'Holes to fill for three decades', *Surveyor* 161 (4723), 6–7.

Reeds, J. (1987) 'Building on landfill – an explosive issue', *Surveyor* 168 (4970), 12–15.

Shimwell, P. (1987) 'Merseyside goes back to the future', *Surveyor* 168 (4975), 12–15.

Sims, F.A. (1984) 'An account of waste management developments in West Yorkshire', *Municipal Engineer* 1 (1), 91–7.

Smith, M.A. (1979) 'Redevelopment of contaminated land: note on the development of landfill sites', *The Engineering Behaviour of Industrial and Urban Fill*, Proceedings of Midland Geotechnical Society Symposium, Birmingham, pp. B49–B70.

Chapter 3

The use of rock, soil and secondary aggregates as landfill cover in South Wales

E.M. Bridges

Safe disposal of the waste created by society has emerged as an issue of considerable significance in the past decade for local authorities who have the responsibility of its disposal. The location and methodology of waste disposal facilities has come under increasingly close scrutiny by the media and an ecologically-aware public. As any form of processing waste adds to the cost of its disposal, most local authorities have resorted to landfill as the least expensive and most convenient option. Disposal of waste in landfill requires a source of suitable cover material to prevent it blowing about in the wind, or being spread around by animals and birds. Covering also reduces smells and restricts the inflow of water into the waste. The survey discussed in this contribution arose from a request for information by the waste disposal authorities of south-west Wales to the Welsh Office which provided funds for the investigation. The work set out to identify the sources of cover material, the range of properties of these materials and the problems of obtaining steady, reliable supplies to cover the wastes as they are dumped.

The present policy for refuse disposal in Britain may be traced back to a report of the Ministry of Health which, in 1932, outlined a number of measures to ensure that domestic refuse was disposed of in as safe a manner as possible. The process of controlled tipping (or sanitary landfill) should be deposited in layers; no layer to be more than a certain thickness (< 2 m); that each layer should be covered on all surfaces exposed to the air with at least 20 cm of earth or other suitable substance; animal, fish or organic refuse should be covered by at least 60 cm; no refuse should be left uncovered for more than 24 hours (Cope *et al.* 1983).

More recently, the effect of the Control of Pollution Act 1974, and the European Economic Community Directives, has been to reinforce these earlier guidelines. Until 1974 advantage had been taken of almost any hole in the ground for tipping of refuse and there is an inheritance of many uncontrolled and potentially dangerous waste disposal sites. Current practice involves the preparation of a Waste Disposal Plan which ensures adequate provision for waste disposal by the authority concerned.

The methods used in the disposal of domestic and commercial waste should reflect the policies outlined in the Department of the Environment's (DoE's) Landfill Practices Review. The most common method of disposal is landfilling, which takes place in cells constructed with earth or other inert material; alternatives are to place the refuse in a large trench or to construct a plateau in layers over the tipping area. The cell method is preferred because the wastes are left uncovered for the minimum amount of time and nuisances such as smell, flies, vermin and scattering of light-weight rubbish by wind are minimized. The embankments of the cellular method conceal the operation from outside the site and when a cell is filled it can be rapidly covered, effectively limiting the infiltration of rainfall into the mass of tipped material. The cell walls may be constructed of other inert materials, not necessarily soil, but care must be taken to avoid seepage through the outside walls where

Table 3.1 Waste collected and estimates of cover material required by waste disposal authorities in south-west Wales

Waste disposal authority	*Waste collected 1984–5 (tonnes)*	*Cover material required 1986 (tonnes)*
Carmarthen	15,816	20,500
Ceredigion	40,000	25,000
Dinefwr	17,500	7,000
Llanelli	42,000	15,000
Lliw Valley	21,000	12,000
Neath	30,100	25,000
Port Talbot	33,790	*
Preseli	20,900	26,400
South Pembrokeshire	16,500	**
Swansea	63,533	85,000

Notes

* Joint disposal with waste from Neath District.

** Joint disposal with waste from Preseli District.

contamination of surface waters may occur. However, when cell walls are made of imported material it represents a loss of space for landfill, and increases operational costs.

Estimates of the amount of waste produced annually by society vary greatly, but in Great Britain as a whole it is estimated to be somewhere between 50 and 100 million tonnes. At the local level, in south-west Wales, which consists of the districts of Carmarthen, Ceredigion, Dinefwr, Llanelli, South Pembrokeshire and Preseli in Dyfed; and Swansea Lliw Valley, Neath and Port Talbot in West Glamorgan, the amount of 'controlled waste' amounted to 527,995 tonnes in 1985–6. Currently, disposal takes place at seventeen sites throughout the area and all these require a reliable, steady supply of cover material for satisfactory site management. As the published figures (CIPFA 1985) are disputed, a survey was conducted in 1986 and the results shown in Table 3.1.

The use of soil or rock materials is of importance at three points in the landfill operation: beneath the tipped material, in the intermediate covering layers and in the final covering system which is constructed to effectively isolate the refuse from its surroundings and to restore the site to an acceptable after-use. Not all materials are suitable for use in all three situations and, where a choice is available, certain materials are preferable. This account reviews the availability of materials for use as cover materials in South Wales.

DEFINITION AND CONCEPT OF A COVERING SYSTEM

A covering system is a sequence of layers of different materials placed over domestic or special wastes in an attempt to isolate them from the rest of the environment. The cover system is an attempt to reduce significantly the infiltration of rainwater into the landfill and also to reduce the amount of leachate released below a landfill. It may be necessary to protect aquifers from this leachate and also to control emissions of gases from the decomposing wastes. If the site is to be reused for any purpose, such as a public open space or playing fields, then it is necessary to design a cover system which will be capable of sustaining plant growth.

Parry and Bell (1985) list the primary functions of covering systems as (a) to prevent exposure of harmful contaminants to human, animal or plant life, (b) to sustain growth of vegetation, and (c) to fulfil an engineering role. These primary functions can be subdivided into several secondary functions which are listed here from the surface downwards and not necessarily in order of importance: erosion control, prevention of dust blowing, support of vegetation, limitation of infiltration, control of soil water movement, control of leachate from the landfill, control of gas movement, prevention of an upward migration of toxic constituents through capillary action, inhibition of root penetration through the barrier, prevention of biological transfer of harmful constituents to the surface, reduction or elimination of harmful surface conditions.

A cover system is constructed primarily as a barrier which separates waste material from life forms to which it may be harmful. Additionally, the cover system is constructed to control or prevent inward infiltration of rainwater and the outward migration of gases, volatile substances and other toxic constituents. To achieve this permeable layers may be added below and above the barrier along which gases and water may flow, conducting them away from the mass of waste. To prevent the rise of contaminants through the cover system a 'break' layer may be incorporated to interrupt the capillary system. This may be provided by gravel or crushed rock aggregate, but where a crushed limestone aggregate is incorporated it can also act as a 'chemical break' which is capable of neutralizing acid solutions and arresting the upward migration of metallic cations.

Few descriptions of cover systems are given in the scientific literature and almost no assessments have been made of their effectiveness as most have been constructed only recently. An example of a cover system for contaminated land is described by McCarthy (1980). It concerns a former domestic refuse landfill site at Hillingdon in London. It was proposed to redevelop it for open space and residential purposes, but it was found to be contaminated with heavy metals. A 150 mm layer of hardcore was placed over the site as a 'break' layer isolating the metallic contamination and providing subsurface drainage; 25 mm of fine porous material was placed upon the hardcore to stop the overlying 150 mm of soil falling down into the break layer. In this example, redevelopment for public open space has proceeded, but the housing development has not taken place. A second example is from Connecticut, USA, where polyvinylchloride (PVC) has been used as an impermeable membrane to isolate a landfill site which was polluting ground waters. A 150 mm sand and gravel layer was placed over the refuse followed by a 100 mm of sand washings to protect the membrane. The membrane extended over the entire landfill and was taken down into a surrounding perimeter ditch. The PVC sheets were joined in the field and a gas venting system installed. Soil material, mixed with sewage sludge and decomposed leaves was laid over the PVC barrier to a depth of 150 mm. After twenty months an improvement in the ground water was seen and monitoring is continuing to observe the long-term success of the cover system (Beck *et al.* 1982).

These two examples illustrate the use of a cover system in an attempt to cure problems created by uncontrolled landfilling in the past. The cover systems for each site have been evolved on empirical grounds and whilst performing well at present, both have yet to be proved satisfactory in the long term. The concepts of cover systems are largely theoretical and monitoring is necessary to assess the success or otherwise of any scheme. When costing a scheme for covering waste, a portion of the budget should be set aside to pay for long-term monitoring.

One site which has been reassessed is described in a survey by JURUE (Joint Unit for Research on the Urban Environment) (1986). This was a 16 ha site which included a sewage treatment works, a domestic refuse incinerator and a waste disposal site. Proposals were made for playing fields on contaminated areas and residential development on uncontaminated land. Soil analysis had indicated contamination by heavy metals. The treatment devised and installed in the early 1970s was a cover of 90–120 cm chalk with 61 cm topsoil on areas to be used as allotments and 30 cm topsoil on recreational areas. An extra layer of rubble was included beneath a children's playground. This cover system was completed but the proposed houses were not built. Subsequently, concern was expressed about the whole site and it was resurveyed in the early 1980s. It was found that the cover system was working satisfactorily ten years after installation.

NATURAL SOURCES OF COVER MATERIAL IN SOUTH WALES

Natural soils and the underlying geological strata are the most readily available materials for use in constructing cover systems. At present they constitute the major source of cover material for waste disposal authorities in South Wales. Although ubiquitous, these materials are not always available for this purpose. Availability is limited by good quality agricultural land, the built-up area, National Parks, and Areas of Outstanding Natural Beauty (AONB) or other restrictions imposed by planning authorities. Sites of Special Scientific Interest (SSSI) and common land also restrict availability of reserves. This review of sources of natural cover materials in South Wales will consider which geological deposits are available for use and also the role soil information can play in identifying the distribution of materials suitable for cover systems.

The geological formations present in South Wales are shown in Figure 3.1. In mid-Wales the rocks are mainly of the Lower Palaeozoic age but in South Wales Upper Palaeozoic rocks occur in the coalfield and smaller areas of Mesozoic strata crop out in the Vale of Glamorgan. In addition, unconsolidated Pleistocene and Holocene deposits are widespread, but of no great thickness. Although geologists classify rocks by time zones, in the present context lithology is more important so this is stressed in the following paragraphs.

Igneous and metamorphic rocks are uncommon in South Wales and exposures occur in restricted areas only, such as the district of Pembrokeshire where the oldest rocks crop out in the St David's peninsula. Extrusive igneous rocks, such as andesites and rhyolites and intrusive igneous rocks such as granite and syenite are present. South of Haverfordwest Pre-Cambrian diorites form the Johnstown ridge. The distribution of these igneous rock outcrops is also shown on Figure 3.1.

The igneous and metamorphic rocks of South Wales are situated on the periphery of the region and in the case of the Pre-Cambrian of Pembrokeshire, they have a limited area of outcrop, and are located in or adjacent to the National Park where restrictions for quarrying limit exploitation. The quarries of Powys are in the Ordovician formation and have the greater reserves remaining to be worked. These rocks are a source of crushed aggregate, but as such they are unlikely to be used simply to cover waste. However, the quarries do have a certain amount of waste which is used as cover material. Where a coarse 'break' layer is required above a toxic waste or contaminated land, or where a conduit for drainage of landfill gas or water is necessary, dolerite or andesite could be used instead of more valuable limestone. Granite or rhyolite also could be used where base-rich properties are not required.

Sandstone, gritstone and conglomerate occur throughout both Lower and Upper Palaeozoic rocks of Wales. In the Cambrian, Ordovician and Silurian formations, these rocks do not make a conspicuous contribution to the landscape as uniform upland plains have been developed rather than the etching out of individual arenaceous strata. In the succeeding Devonian, sandstones occur

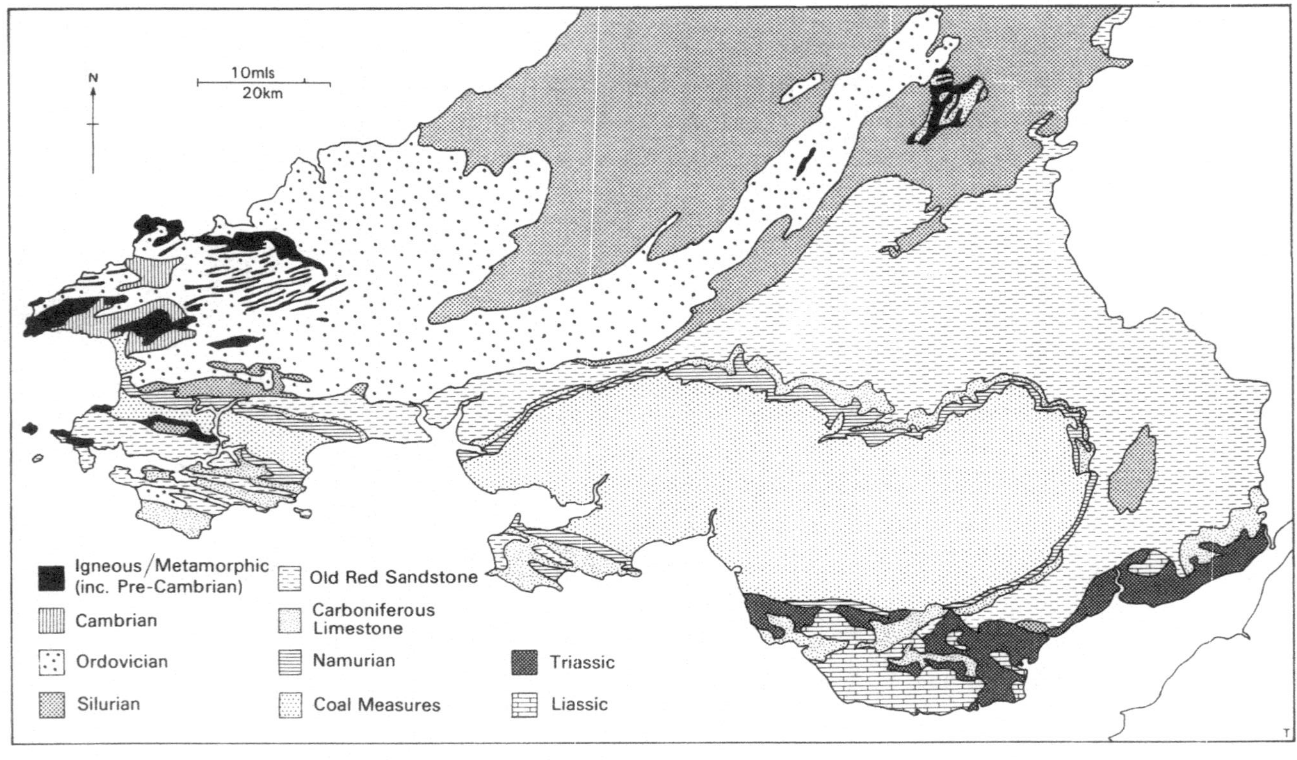

Figure 3.1 General geology map of South Wales showing location of igneous and metamorphic rocks

more significantly higher in the sequence, where they form the northward-facing escarpment of the Brecon Beacons and Carmarthen Fans where elevations of almost 1,000 m are reached. Thin conglomerates, called the Plateau Beds, cap the highest point of the Brecon Beacons. Coarse-grained conglomerates are also a feature of the Old Red Sandstones in Pembrokeshire, Gower and the Vale of Glamorgan.

The Coal Measures are characterized by a repeated sequence of sandstone, shale and coal strata, but in the South Wales coalfield the upper part of the Middle Coal Measures is dominated by the Pennant Sandstones which form the plateau surface into which the South Wales valleys are cut. The Pennant Sandstones are dark grey when fresh but weather to an ochreous brown colour. Thin shale partings between the sandstones provide sufficient clay to give a loamy soil. Sandstones of the Triassic rocks in South Wales are restricted in area but near Bridgend, the Quarella sandstone is a significant arenaceous facies.

These sandstones, gritstones and conglomerates represent a large resource of aggregate and potential cover material which is widely distributed throughout South Wales. Many small quarries were worked in the last century and a few siliceous sandstones continue to be quarried for refractory purposes. In the counties of Dyfed and Powys, permitted reserves of 145.8 million tonnes are available from Pre-Cambrian and Lower Palaeozoic rocks. In the coalfield counties of Gwent, Mid- and West Glamorgan the available resources from the Pennant Sandstones are 21.8 million tonnes. These figures refer to areas which have been granted planning permission for quarrying of aggregates. Although the aggregate itself may be too expensive, there is up to 20 per cent waste resulting from the crushing process which is a valuable source of cover material for waste disposal authorities.

Restrictions upon the use of sandstone throughout South Wales for aggregate and cover material would include the limitation of quarrying in the Brecon Beacons National Park and the impact modern quarrying may have on other visually sensitive areas. As the various sandstones usually form prominent landscape features, exploitation is difficult without loss of visual amenity. With a heritage of despoiled land, few local authorities in South Wales would wish to create more disturbed land in their search for cover material.

Shales and mudstones of the Ordovician and Silurian formations form an extensive outcrop in the northern part of South Wales. These rocks are usually grey, interbedded with thin sandstones and occasional igneous intrusions. In some places these argillaceous rocks are sufficiently metamorphosed to be slates. The lower Devonian comprises red-coloured mudstones with subsidiary siltstones or sandstones. Some thin limestones are also characteristic of these strata, especially in Gwent. In the Carboniferous sequence of South Wales, grey shales predominate in the Millstone Grit and the Lower Coal Measures, but the geographical extent of the Pennant Sandstones and superficial materials limits their outcrop. Red mudstones are characteristic of the Triassic rocks in the Cardiff district.

The widespread occurrence of argillaceous rocks in South Wales suggests that they might form a useful source of cover material should other supplies become scarce. The Lower Palaeozoic rocks tend to break into tabular shale fragments but crush easily to a finer aggregate. However, many of the mudstones may be difficult to work in wet weather but, where weathered, they may provide a reasonable cover if restriction of rainfall percolation is required above wastes.

Clays are normally quarried for brick-making and slate for roofing material. Former brickworks quarries may have considerable amounts of reject material depending upon the quality of the clay being worked. Slate quarrying generates large quantities of waste. In South Wales alone, 50 million tonnes of slate waste are reported to exist: this could be a possible source for cover material in locations near to the Preseli Hills and Llangynog in Powys. Coal-mining has resulted in tips of colliery waste associated with all former mines. Many tips have been landscaped and made safe but 50 million tonnes remains available for other uses, one of which may be as cover material.

Two limestone formations crop out in South Wales, the Carboniferous Limestone and the Liassic Limestone. The former occurs around the coalfield and the latter only in the Vale of Glamorgan. Numerous quarries have exploited the Carboniferous Limestone for use as building stone, as a flux in the iron and steel industry, as road metal and for lime and cement. Many quarries on the northern rim of the coalfield have closed but several still operate in the Vale of Glamorgan. The Liassic Limestones are thinly bedded with shale partings; these shales decrease in importance westwards, but at Aberthaw are in the correct proportion for cement manufacture.

Crushed limestone aggregate reserves from the Carboniferous Limestone outcrops in the Vale of Glamorgan amount to 251.9 million tonnes. However, there is strong demand for limestone aggregate and it is used for many purposes where a less valuable rock could be substituted. Limestone is unlikely to be used for cover material unless a calcareous break layer is required in a specially designed cover system for contaminated areas, but 'scalpings' are available from most limestone quarries which are used for cover materials.

Both Carboniferous and Liassic Limestones could be sources of material which would be ideal where a calcareous cover material is desired. However, the Liassic Limestones lie beneath some of the most fertile land in the Vale of Glamorgan and abutt against a Heritage Coast and so are unlikely to be much exploited for cover material. Many lowland areas of the Carboniferous Limestone also bear fertile soils or are included in landscapes of high scenic quality as in the Gower Area of Outstanding Natural Beauty. These restrictions on quarrying greatly limit the available areas from which limestone may be obtained.

Superficial materials lie upon the surface of the solid rocks discussed previously. These include gravels, sands, tills (boulder clays), lacustrine and alluvial deposits. Their thickness and distribution is very variable. They have been mapped by the Geological Survey in South Wales, but north of the Brecon Beacons no recent geological surveys have been published. So, the only information available there is from surveys at the end of the nineteenth century or the beginning of the present century.

Two major periods of glaciation are known to have affected South Wales, those attributed to the Devonian being the most significant. Areas outside the limits of the Devonian glaciation have subdued remnants of an earlier, more widespread, (Wolstonian?) glaciation. Many of these glacial deposits have been affected by periglacial activity and some have been covered by a thin aeolian deposit. The glacial deposits may be considered under the subheadings of sands and gravels, and tills.

Sands and gravels of glacio-fluvial origin are uncommon in South Wales and contribute little to the aggregate industry. Their main area of occurrence is in the Vale of Glamorgan lying between Cardiff and Bridgend. Smaller areas of similar deposits lie in Singleton Park and Sketty in Swansea and at Margam. Elsewhere in South Wales sand and gravel deposits are confined to narrow strips alongside

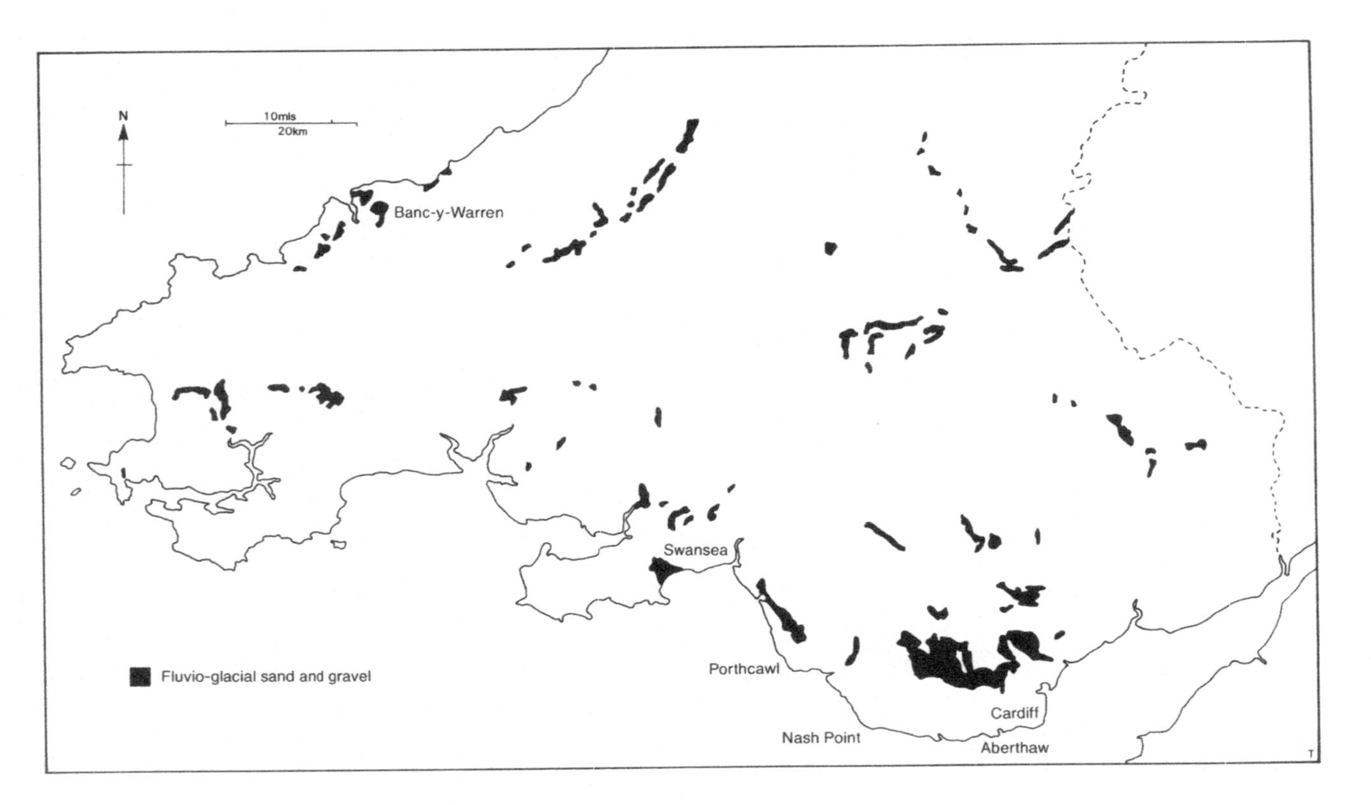

Figure 3.2 Distribution of fluvio-glacial sand and gravel in South Wales

rivers and the occasional kame feature as at Banc-y-warren, Dyfed (see Figure 3.2).

These gravels are very poorly graded with a large range of particle sizes and they are often mantled by a layer of finer material of aeolian origin. Consequently they have been little exploited by the sand and gravel industry. It is estimated that there are 1.7 million tonnes of land-based sand and gravel reserves, but these are supplemented annually by approximately 1.5 million tonnes of marine-dredged sands and gravel from the Bristol Channel. Between Gower and the North Devon coast, at depths of < 50 m, a thin (< 0.5 m) layer of sediment overlies bedrock. This material on the sea floor consists of sandy gravel or gravel with areas of bedrock exposed. At present, sand extraction takes place in shallow seas over sand banks which lie west of Nash Point and Porthcawl and, to a lesser extent, Gower. No reliable estimates are available of these marine resources, but it is possible to extract sand and gravel from depths of 25 m (70–80 feet), although local dredgers normally operate at 10 m (30 feet).

As a potential cover material the glacio-fluvial sands and gravels could be a valuable resource in the development of cover systems. Even after the removal of gravel, the wide particle-size range suggests it would make a useful cover material. The marine sands and gravels are less likely to be used for cover material as the cost of extraction restricts their use to building aggregate.

In the Pleistocene glaciers moved slowly across the landscape and rock material was subject to attrition; it was eroded and eventually deposited as an assorted mixture of boulders and finer material, referred to as till (see Figure 3.3). The nature of these deposits reflects the geological formations over which the ice passed. Tills derived from the Pennant Sandstone outcrops tend to be sandy, but material from the Lower Palaeozoic mudstones is clayey and blue-grey in colour. These tills have been deposited as glacial ice melted, so they are typically unstratified, but occasionally some stratification by melt-water may be observed.

Periglacial activity following deglaciation has superficially reworked most of these tills. In the coalfield these superficial materials occupy the middle and lower slopes of many valley sides where they are the site of landsliding activity. Till deposits are widespread on the plateau surfaces between the South Wales valleys, often reaching a thickness of several metres. It is impossible to generalize about the composition and physical properties of these

Figure 3.3 Distribution of boulder clay (till) in South Wales

materials as they can range from fine clays to gravels. Their usefulness as potential cover materials will depend very much on the local variations in composition. The distribution of till throughout South Wales is shown in Figure 3.3, but north of the coalfield the information available is limited and its extent is underestimated.

Where these materials are encountered in road reconstruction or other major civil engineering projects they can form an extremely good source of cover material. In future, should there be a shortfall in the amounts of cover required, it would be desirable for engineers to consider planning the construction of major projects to a lower datum to release greater supplies of material. Perhaps it should be made mandatory for promoters of major works programmes to supply WDAs with suitable geological materials.

The river valleys of South Wales all contain deposits of water-sorted material. Fluvial processes often result in a gravelly basal layer with finer silty deposits making up the floodplain alongside a river. This sort of section may be observed inland where alluvial deposits are only 1–2 m in thickness. However, near the coast, the rivers valleys of south Wales have been infilled with up to 48 m of alluvial or estuarine material in response to the rising sea-level of post-glacial times. Investigations in the Neath and Tawe valleys have revealed deep infillings which occupy an irregular floor with ice-scoured rock basins beneath the alluvium. In the Loughor estuary, till occurs beneath calcareous silty alluvia which are interbedded with layers of peat.

The coastal regions of South Wales have considerable accumulations of quartzose sands, at Merthyr Mawr, Margam, Pennard, Oxwich, Broughton Burrows and Pendine, with smaller deposits elsewhere. These sands are probably derived from outwash deposits of the Devensian glaciers. The action of waves and wind during post-glacial time has gradually pushed these sands landwards until they assumed their present positions. Coniferous trees have been planted to stabilize the dunes at Pembrey and these coastal areas have a high priority for recreational activity. Some sand extraction took place in the past, but it is now generally agreed that the supply of sand from the sea floor is limited and that its removal from the dunes is undesirable and should be discontinued. For amenity reasons it is unlikely that these sands will ever provide a source of cover material.

Many upland areas of Wales are thinly covered by blanket peat the growth of which was encouraged during the wetter Atlantic and Sub-Atlantic periods of post-glacial time. This peat is strongly acid and highly retentive of water. Although some peats were cut for fuel in the past, they are areas of great biological interest and unlikely to be used as cover material.

THE USE OF PEDOLOGICAL INFORMATION IN THE SEARCH FOR COVER MATERIAL

The Soil Survey of England and Wales has recently compiled maps of the soil resources of the whole country including Wales (Sheet 2) at a scale of 1:250,000. An accompanying bulletin describes the mapping units (Rudeforth *et al.* 1984). Greater detail is given in surveys of part of the Vale of Glamorgan (Crampton 1972) and 1:25,000 sheets SM90/91 (Rudeforth 1974), SN41 (Clayden and Evans 1974), SN13 (Bradley 1976), SN24 (Bradley 1980), SN62 (Wright 1980), SN72 (Wright 1981). Other information is available in scientific journals (Crampton 1961, 1966; Bridges and Clayden 1971; Bridges 1975, 1985).

Soils are distinguished from each other by their profiles, maturity, texture, structure, stone content of the horizons present and the mode of origin of the parent material. It is this last criterion which makes the maps of the pedologists useful in the search for cover material. Although the maps have been made primarily for agriculturalists whose interest is in the soil as a growth medium, consideration must be taken of the subsoil properties which may affect crop performance. Soil is an extremely valuable resource which is hardly likely to be used for its incidental properties. Topsoil is much too valuable for use as a cover material, but subsoil and parent material could be made available for covering wastes where land is developed for non-agricultural purposes.

Soil associations in South Wales

The most common soil types which occur in South Wales, called subgroups by Avery (1980) are:

Typical Brown Earths	Cambic Stagnogley Soils
Gleyic Brown Earths	Cambic Stagnohumic Gley Soils
Typical Argillic Brown Earths	Ferric Stagnopodzols
Typical Brown Podzolic Soils	

and smaller areas of Rankers, Alluvial Soils and Sand Pararendzinas.

Typical Brown Earths are non-calcareous soils with profiles which have an 'altered' cambic B horizon, lack gley features and are not formed in alluvium. *Gleyic Brown Earths* have similar properties but have evidence of gleying as indicated by mottling. *Typical Argillic Brown Earths* have an argillic B horizon enriched with eluviated silicate clays and lack gley features. *Typical Brown Podzolic Soils* have a podzolic B horizon enriched with iron, aluminium and humus but lack a subsurface bleached horizon. *Cambic Stagnogley Soils* have poor drainage as indicated by mottling and the B horizon is 'altered' rather than enriched with iron, clay or humus. *Cambic Stagnohumic Gley Soils* are similar but have an organic peaty surface horizon. *Ferric Stagnopodzols* have a peaty topsoil and a bleached subsurface horizon overlying an iron-enriched subsoil. *Brown Rankers* are shallow non-calcareous soils, *Alluvial Gley Soils* are poorly-drained soils influenced by a high water table and *Sand Pararendzinas* are developed in Blown Sands containing calcareous shell fragments.

The 1:250,000 Survey by the Soil Survey of England and Wales gives a detailed impression of the major soil associations and their distribution is shown in a simplified map (Figure 3.4). The soils most suited for use as cover materials are included in the Arrow, Brickfield, Milford, and Wilcocks associations, the characteristics of which are briefly outlined in the following paragraphs.

Gleyic Brown Earths of the Arrow Association are formed in glacio-fluvial sands and gravels which occur in the Vale of Glamorgan, north of Cowbridge. These soils are coarse loamy in texture and are developed on hummocky topography where drainage conditions are very variable. The depth of the parent material overlying the solid strata is unknown but is also likely to be variable. Currently under poor pasture or forestry (Hensol Forest) these areas could provide a source of useful cover material. Soil profiles described by Crampton (1972) indicate that they often contain indurated subsoils where the soil material has been compacted naturally by periglacial processes.

Cambic Stagnogley soils developed in glacial till or head derived from Carboniferous shales and sandstones dominate the Brickfield Association. These soils are extensive north and west of Swansea where they are often associated with the more organic soils of the Wilcocks Association. Brickfield soils are waterlogged seasonally and so are normally under pasture or forestry. The loamy texture of the parent material continues to depth and it includes stones of mixed provenance. The depth of the deposit is not known in detail and it is

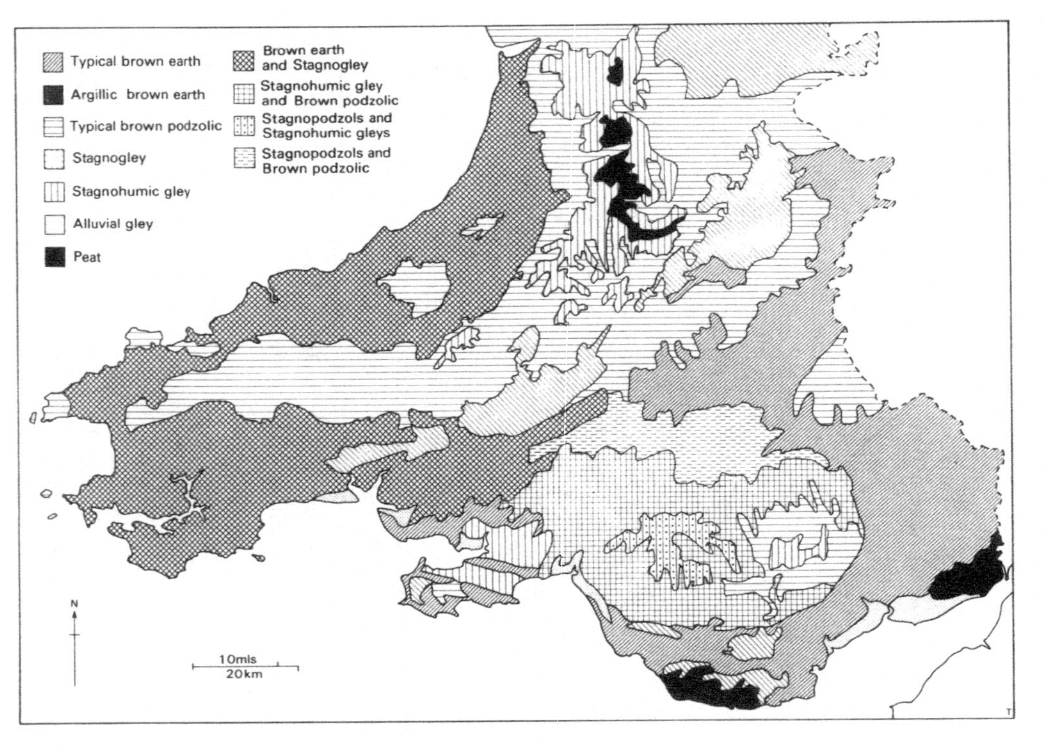

Figure 3.4 Distribution of major soil groups in South Wales

likely to be very variable. Soils in this grouping have a wide range of particle size and so could provide a useful source of cover material as they can be compacted tightly.

The Milford Association is an extensive grouping of soils on the outcrop of the Devonian marls in South Wales from Milford Haven to the English border. The soils are reddish-brown stony clay loams, of medium depth and are typical Brown Earths. In themselves, these soils do not possess much potential for use as cover material, but the parent material is a relatively soft mudstone and could easily be excavated. The wide extent of the outcrop of the marls suggests this material may be a resource of some significance.

Strongly-gleyed soils with peaty topsoils of the Wilcocks Association are developed from grey glacial drift deposits derived originally from Lower Palaeozoic and Carboniferous rocks and are common on the gently-sloping uplands of South Wales. Similar soils on reddish drift have been mapped as the Wenallt Association. These soils are normally about 100 cm in depth and the parent material considerably deeper, although the contour of the underlying solid rock is generally unknown. In certain circumstances these soils could provide a significant source of cover material. However, their fertility is low and adequate preparatory study would be necessary before exploiting them.

Many of these soil associations contain areas of shallow soils which are unlikely to contribute greatly to reserves of cover material. Other limitations include the large areas devoted to agriculture or built upon and so are not available for use as cover material. By common consent many areas will not be considered to be available as they occur in National Parks or Areas of Outstanding Natural Beauty. The areas which remain might be thought of as possessing potential as sources of cover material. Consequently, when any constructional activity is planned in the areas where these soils occur, arrangements could be written into the planning consent for the supply of suitable material to the waste disposal authority.

SECONDARY AGGREGATES AS COVER MATERIAL

The use of rock and natural soils for cover material unfortunately necessitates digging up one part of the countryside to reinstate another. Consequently if suitable alternative materials are available to cover wastes, these should be seriously considered as substitutes and used whenever possible. In the long term it may be desirable to make these materials available by statute, if necessary.

The staff of the Building Research Establishment have studied the possibilities of the use of major industrial by-products and waste materials (Gutt *et al.* 1974), but do not cite cover material as a useful means of their disposal. In South Wales, the South Wales Working Party on Aggregates (1977–83) produced a report in 1977, followed by a supplement in 1979, regional commentaries in 1980 and 1981 and a second supplement in 1983. Although these documents are concerned mainly with the supply of natural aggregates for the construction industry, they also consider the use of 'secondary aggregates' derived from industrial waste products. A guide to the use of secondary aggregates has been produced by the British Standards Institute (BSI 6543: 1985). This guide reports upon the arisings and disposal of colliery spoil, pulverized fuel ash, furnace-bottom ash, quarry wastes, incineration wastes, metallurgical slags, red mud, china clay wastes, spent oil shale, calcium sulphate, phosphogypsum and demolition wastes. Their location in South Wales is shown in Figure 3.5.

Mechanized coal-mining techniques have increased the quantity of unsaleable waste material brought to the surface with the coal. Two hundred years of mining in the coalfields of Britain have left a stockpile of over 3,000 million tonnes. Current production of colliery waste amounts to about 60 million tonnes a year. Many former colliery waste heaps have been made safe and landscaped during the past twenty years, but the quantities which remain form a significant resource for cover materials. In 1981 the South Wales Working Party on Aggregates recorded that 6 million tonnes of colliery waste was produced annually in South Wales, of which only 0.75 million tonnes was sold for constructional fill. Total resources at that date were estimated at 50 million tonnes.

Coal shale waste is not hazardous but the pyrites in it gives a strongly acidic weathering product and some shales release soluble salts on weathering. The surface layer of old tips has mostly completed these weathering changes but lower layers may still be capable of releasing acidity. This material can be used for restoration purposes but precautions must be taken to counter the excessive acidity. However, coal shale, being a mudstone, tends to form a slurry when subjected to traffic and is not favoured by disposal site operators.

Quarries have been opened for many purposes connected with the construction industry in a wide range of different rock types.

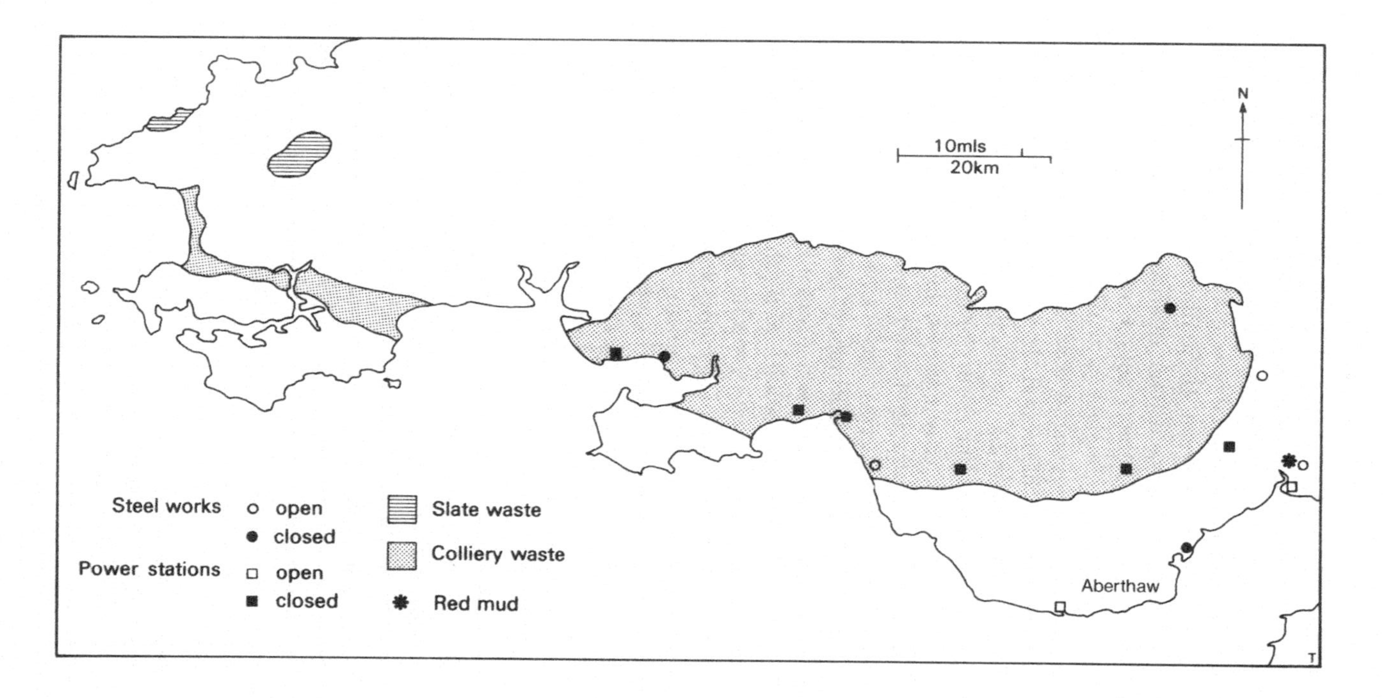

Figure 3.5 Sources of secondary aggregates in South Wales (Gutt *et al*, 1974; South Wales Working Party, 1977–83; British Standards Institute, 1985

Specific quarries will have an output of a particular type of stone or aggregate but will also create some reject material which is tipped alongside the quarry or in a worked-out area. This waste material may include the overburden of soil or weathered rock and is usually suitable for use as cover material. In South Wales there are several limestone quarries, some sandstone quarries and a few gravel workings, all of which can be a source of cover material. One historical waste material is the slate waste which accumulated in the Preseli district and at Llangynog in Powys. As amounts of waste are 20–30 times the quantity of dressed slate produced, quantities are large, and an estimated 50 million tons exists in South Wales alone. However, coarse-grained material does not cover effectively, leaving gaps where flies breed, so crushing to a finer grain size may be desirable. In this respect, dredged sand and silt from harbours, canals and river mouths may prove useful cover material if it is free from toxic material (Bramley and Rimmer 1988).

Pulverized fuel ash (pfa) results when pulverized coal is burnt in a stream of air in modern power-stations. This ash comprises particles of fine sand size (60 per cent) with a specific gravity of between 1.9 and 2.4, but about 5 per cent of it consists of very light, hollow cenospheres with an apparent specific gravity of 0.5. Compared with soil materials, pfa lacks a clay fraction and nitrogen, it is alkaline (pH 11–12) and contains soluble salts. Lagooning helps to reduce the soluble salts but an additional problem is the presence of between 3 and 250 mg/kg^{-1} water-soluble boron. It is generally accepted that 4–10 mg/kg^{-1} boron in soil is slightly phytotoxic, so the presence of boron in pfa may present some problems. However, the content of boron in South Wales coals is lower than elsewhere in Britain. The pfa is difficult to handle as it is easily blown about by the wind, and it will flow like a liquid if aerated. Consequently, it is usually handled as a slurry in a 60:40 mixture with water. It is discharged into lagoons to sediment so that the water may be used again. Ultimately it is dug out of the lagoons and dumped where its pozzuolanic properties cause it to harden as it dries. Despite these difficulties, pfa is a valuable material which could play a greater part in waste material management. Anther product of power stations is furnace-bottom ash (fba) which accumulates in the base of the furnaces where it fuses to produce a porous clinker-like material.

The number of power-stations in South Wales has been reduced from seven which were operative in 1970 to the three (on two sites) which are sufficient to supply South Wales today. Some pfa may be present on former power-station sites (for example the former Tir John Power Station site in Swansea). The production of pfa + fba in the region in 1981 was 0.772 million tonnes and the stockpile of these materials is 4.5 million tonnes. Only about 14 per cent of current production is used, which leaves considerable quantities available for use as cover material.

The use of metalliferous slags for fill material on construction sites has occurred for many years. Broadly, it is possible to divide these materials into those from iron- and steel-making and those from non-ferrous metal smelting. The former have wide use in industry and agriculture, but the latter still contain sufficient toxic material to make them difficult to use. Blast-furnace slag is relatively homogeneous when tapped from the furnace but it is cooled in three ways: cooled in the air it sets into a rock-like consistency, when cooled with a limited amount of water it forms a granular material, and when cooled with excess water a glassy granular material is formed. Blast-furnaces are situated at Llanwern and Port Talbot in South Wales and in 1981 these two works produced 900,000 tonnes of slag. Blast-furnace slag is composed of silicates and alumino-silicates of lime and other bases. It is used as roadstone, railway ballast, as a raw material for cement, and as a filter medium for sewage disposal. It has not been used as a cover material, although there is the possibility that it could be used in association with colliery shale to neutralize acidity caused by weathering of pyrites. Resources of blast-furnace slag in South Wales include the wastes of the former iron industry at Merthyr Tydfil and Ebbw Vale where works have closed; these are being worked for constructional fill.

Current production of steel slag comes from the basic oxygen process or electric-arc furnaces at Llanwern and Port Talbot works. These slags may contain lime and magnesia which might make them unstable so they should be weathered before use as fill under buildings or roads or as aggregate for concrete. The total production of steel slag in South Wales in 1981 was 728,000 tonnes, most of which was used in South Wales for constructional fill with smaller quantities going for roadstone, rail ballast and concrete aggregate. Only a small quantity was stockpiled.

Virtually all of the non-ferrous metal slags of the Swansea Valley have now been sandwiched between iron and steel slag to make a new, level constructional surface on the valley floor. As a resource for cover material these are no longer available and in any case their metal content would have made them suspect. The only other metalliferous waste known to be present in South Wales are deposits of red mud from alumina refining. Gutt *et al.* (1974) draw attention to 'many tens of thousands of tonnes' of red mud at Newport, Gwent.

Incineration of refuse decreases its bulk by 90 per cent which has great attractions for local authorities and for the general public who are reluctant to see extensive landfills, even if well sited by the planning system and well managed. Problems have been encountered in waste incineration as the residue of the incinerators may contain potentially toxic metallic elements and the waste gases, dioxins. There are some incineration plants in South Wales but no information is to hand of the use of their residues for cover material even though Gutt *et al.* (1974) state that 'it is known to be used in part for the covering of tips of raw refuse'. There are 2 million tonnes produced annually in England and Wales, but it is unimportant in South Wales.

Concrete and rubble from buildings, roads, pavements, airfield runways and parking aprons may be useful for cover material in certain circumstances, but contamination with wood, glass, gypsum plaster, asphalt and steel reinforcing rods can detract from the usefulness of this material. The other sources of secondary aggregate mentioned in the national reports, china clay waste, phosphogypsum and calcium sulphate do not occur in South Wales and are not considered here as transport costs to South Wales would preclude their use as cover material.

CONCLUSIONS

The search for cover material should begin on-site with investigations to see if local materials are suitable for use in a cover system, as a layer of restricted permeability or as a support for vegetation on its surface. Loamy soil materials are preferable where compaction is required to decrease permeability, and more effective compaction is achieved on the wet side of the optimum moisture content. It is also necessary to select the soil with a type of clay which will perform satisfactorily under the prevailing climatic conditions.

Waste material from the aggregate industry has made an important contribution to the waste disposal authorities' requirements for cover materials. The aggregates themselves could be used, but the expense of processing the rock makes this unlikely except for specific purposes, such as the emplacement of a layer of aggregate as a break layer, or as a layer where gases or water are allowed to seep away from the landfill. There has been a tendency for authorities to use crushed limestone, but in many cases another rock type would be equally suitable. Only where a high pH barrier is required between the waste and the environment is it essential to use the scarcer, more expensive, limestone. It has been identified that waste material from sandstone quarrying activity is under-used as is the coal shale waste from the mining industry throughout the South Wales coalfield. In Preseli and in the extreme north of Ceredigion slate waste is available in considerable quantities. It would certainly be possible to use this material to cover wastes but it would be expensive to transport it to the waste disposal facilities in the more populous areas. The slate waste would be satisfactory as a break layer, or for conduction of water, but would not be effective if an impervious cover system was required.

As a result of the work of the Soil Survey of England and Wales, the distribution of Welsh soils is now known with some accuracy, but detailed information about the physical properties of the soils is not available. This is an area where further information is urgently required; soil materials should be used in ways so that their physical properties are used for environmental protection. It is no longer acceptable simply to use them to cover a landfill with no thought of what else might be achieved by the material in a properly designed cover system. Parameters such as particle size, bulk density, liquid and plastic limits, compactability, shear strength and type of clay mineral present are important factors to take into consideration. Guidance on the use of soils for landfill cover is given by the National Soils Handbook (USDA 1983).

The use of industrial by-products for cover materials is obviously to be encouraged and all authorities have preferential terms for accepting inert wastes which can be utilized. Slags from the smelting of copper, lead, zinc and other 'heavy' metals should be avoided as they normally contain some residue of these toxic metallic elements. Blast-furnace slags usually have high pH values and can be used in many situations but stability problems may arise if used where buildings are to be constructed. Large quantities of blast-furnace slags are not now produced, and the alternative demands for them will probably take precedence over their use as cover materials.

Analysis of the situation in south-west Wales has revealed that the rather haphazard methods of cover material supply which operated in the past are no longer adequate to meet present-day requirements. The supply of suitable cover materials is a problem with which all waste disposal authorities must contend and so the relevance of an assessment of the waste disposal authority's requirements is obvious. In detail, however, different types of cover material are needed as landfilling proceeds, so it is advisable for the authorities concerned to have an inventory of the sources of supply as well as of the type of material available. In this way the most effective use is made of the cover materials present.

ACKNOWLEDGEMENTS

The financial support of the Welsh Office in the preparation of Contract WEP126/100/2 is gratefully acknowledged as is the active participation of Messrs. D.J. Leech and A.T. Evans in the work of the project. Thanks are also due to Mr Guy Lewis, cartographer at University College, Swansea.

REFERENCES

Avery, B.W. (1980) *Soil Classification for England and Wales*, Technical Monograph no. 14, Harpenden: Soil Survey of England and Wales.

Beck, W.W., Dunn, A.L. and Emrich, G.H. (1982) 'Leachate quality improvement after top sealing', in D.W. Schultz (ed.) *Proc. 8th Annual Research Symposium on Land Disposal of Hazardous Wastes*, pp. 464–74, Ft Mitchell, Kentucky.

Bradley, R.I. (1976) *Soils in Dyfed III Sheet SN13 (Eglwyswrw)*, Soil Survey Record no. 38, Harpenden: Soil Survey of England and Wales.

Bradley, R.I. (1980) *Soils in Dyfed V Sheet SN24 (Llechryd)*, Soil Survey Record no. 63, Harpenden: Soil Survey of England and Wales.

Bramley, R.G.V. and Rimmer, D.L. (1988) 'Dredged materials – problems associated with their use on land', *Journal of Soil Science* 39, 469–82.

Bridges, E.M. (1975) 'Soils in parent materials formed from Devonian rocks in Wales', *Welsh Soils Discussion Group Report* 15, 73–93.

Bridges, E.M. (1985) 'Soil survey in Gower', *Welsh Soils Discussion Group Report* 26 (in press).

Bridges, E.M. and Clayden, B. (1971) 'Pedology', in W.G.V. Balchin (ed.) *Swansea and its Region*, Swansea: British Association.

British Standards Institute (1985) *Use of Industrial By-products and Waste Materials in Building and Civil Engineering*, BSI 6543:1985.

Chartered Institute of Public Finance and Accountancy (CIPFA) (1985) *Waste Disposal Statistics 1984–1985 Actuals and Estimates*, Statistical Information Service.

Clayden, B. and Evans, G.D. (1974) *Soils in Dyfed I Sheet SN41 (Llangendeirne)*, Soil Survey Record no. 20, Harpenden: Soil Survey of England and Wales.

Cope, C.B., Fuller, W.H. and Willets, S.L. (1983) *The Scientific Management of Hazardous Waste*, Cambridge: Cambridge University Press.
Crampton, C.B. (1961) 'An interpretation of the micro-mineralogy of certain Glamorgan soils: the influence of ice and wind', *Journal of Soil Science* 12, 158–71.
Crampton, C.B. (1966) 'Certain effects of glacial events in the Vale of Glamorgan', *Journal of Glaciology* 6, 261–6.
Crampton, C.B. (1972) *Soils in the Vale of Glamorgan*, Harpenden: Memoir of the Soil Survey of England and Wales.
Gutt, W., Nixon, P.J., Smith, M.A., Harrison, W.H. and Russell, H.D. (1974) *A Survey of the Locations, Disposal and Prospective Uses of the Major Industrial By-Products and Waste Materials*, Current Paper CP19/74, Building Research Establishment, DoE.
JURUE (1986) *The Reuse of Landfill Sites*, Birmingham: Ecotec.
McCarthy, M.J. (1980) 'Reclamation of a refuse tip for open space and housing development', *Proc. Conference on Reclamation of Contaminated Land*, Paper B8, London: Society of Chemical Industry.
Parry, G.D.R. and Bell, R.M. (1985) 'Covering systems', in M.A. Smith (ed.) *Contaminated Land: Reclamation and Treatment*, Plenum: NATO-CCMS (Committee on the Challenges for Modern Society).
Rudeforth, C.C. (1974) *Soils in Dyfed II Sheets SM90/91 (Pembroke/Haverfordwest)* Soil Survey Record no. 24, Harpenden: Soil Survey of England and Wales.
Rudeforth, C.C., Hartnup, R., Lea, J.W., Thompson, T.R.E. and Wright, P.S. (1984) *Soils and their Use in Wales*, Bulletin no. 11, Harpenden: Soil Survey of England and Wales.
South Wales Working Party on Aggregates *Interim Report* (1977); *Supplement to Interim Report* (1979); *Regional Commentary Part 1* (1980); *Regional Commentary Part 2* (1981); *1981 Aggregate Minerals Survey* (1983); Cardiff: Mid-Glamorgan County Council.
USDA (1983) *National Soils Handbook*, United States Department of Agriculture, Washington.
Wright, P.S. (1980) *Soils in Dyfed IV Sheet SN62 (Llandeilo)*, Harpenden: Soil Survey of England and Wales.
Wright, P.S. (1981) *Soils in Dyfed VI Sheet SN72 (Llangadog)*, Harpenden: Soil Survey of England and Wales.

Chapter 4

The use of special waste consignment note data in waste planning for the Greater London area

Julian Parfitt

This chapter examines the potential for using data generated by existing legislation to learn more about hazardous wastes in the Greater London area – its composition, to where it goes, and how disposal and treatment options are spatially distributed. The main features of waste regulation and planning are described, and the relationship between the administration of regulations and intelligence gathering is examined for London. Overall trends in London's hazardous waste arisings are presented, followed by a detailed analysis of special waste consignment notes.

BACKGROUND TO POLICY, LEGISLATION AND THE WASTE DATA PROBLEMS

Specific controls over waste management have been a relatively recent addition to pollution control legislation, and the waste planning requirements of the Control of Pollution Act have been in effect for barely a decade. In preparing waste plans, waste disposal authorities are expected to assess options for their area such as landfill, incineration, waste reduction, recycling, and waste treatment (Control of Pollution Act 1974: Part 1, Section 2).

Waste planning and resource allocation within waste regulatory agencies should be based on sound data; so far the quality of data collected by many waste disposal authorities for waste disposal plans have been poor, particularly for industrial wastes. The Hazardous Waste Inspectorate commented in their third and final report that: 'Unfortunately the needs of disposing of industrial waste, in many instances, are rather vaguely stated and generally left to market forces to resolve' (Hazardous Waste Inspectorate 1988: 50). An important contribution to this ignorance is the way in which waste regulation operates in relation to waste life cycles.

The emphasis of British waste regulation rests at the 'very back end' of the waste life cycle: at the point where wastes arrive at a licensed disposal site (Wynne 1987: 197). This makes the task of relating waste arisings back to particular processes, or of defining waste compositions, more difficult. In addition, dependence on the policy of 'co-disposal' (whereby hazardous wastes are deposited in secure landfill sites mixed in with domestic refuse), requires minimal information about waste composition compared with the type of knowledge needed for other options, such as incineration or resource reclamation. This factor can only further reduce the quality of information on hazardous waste streams, as well as hindering independent assessment of 'best practicable means' for their treatment or disposal.

Potentially there exists a source of detailed information for the more hazardous wastes from the two pieces of legislation that have been enacted since 1973 to regulate their movements. The first was the Deposit of Poisonous Wastes Act 1973, which was replaced in March 1981 by the Control of Pollution (Special Waste) Regulations 1980.

The Deposit of Poisonous Wastes Act (DPWA) regulated a more extensive group of hazardous wastes than the Special Waste Regulations, based on an exclusive list principle which was fail safe. Whereas the Special Waste Regulations control wastes which might pose an immediate threat to human life, the DPWA also covered wastes that pose a threat to the environment. The Department of the Environment has argued that the environmental hazard category is now controlled by the site-licensing provisions of the Control of Pollution Act 1974, through the consideration of the 'difficult wastes' list (Royal Commission on Environmental Pollution 1985: 13). However, the Hazardous Waste Inspectorate (now a part of Her Majesty's Inspectorate of Pollution) concluded, as a result of a large number of site visits, that 'it is legitimate to question the value of many licenses in terms of environmental protection' (Hazardous Waste Inspectorate 1986: 55).

A principle criticism of the DPWA was that many waste disposal transactions were notified that need not have been. The new legislation was designed to relieve waste disposal authorities and waste producers of unnecessary paperwork and to enable the regulators to concentrate more on field control. However, the total number of consignment notes generated was higher than expected and a season-ticket scheme for the most frequent producers of special wastes was introduced (Department of the Environment 1985: Annex 4).

Under the Control of Pollution (Special Waste) Regulations 1980, six copies of a standard consignment note form are involved in the various stages between waste generation and disposal. At least three days prior to dispatching a consignment of special waste, the producer must send a pre-notification consignment note copy with details of waste composition to the 'away' waste disposal authority (the one in whose area the site due to receive the waste is situated). On collection of the consignment by the carrier the name and date of collection are recorded on another copy of the consignment note, and a further section is completed by the producer. This certifies that the correct information about the consignment has been declared and that advice on 'appropriate precautionary measures' was given to the carrier. in the final stage, the treatment or disposal site operator verifies that the wastes described by the producer can be received at that site under its site-licensing conditions. This copy is then sent back to the waste producer's waste disposal authority who then match it to their original copy, thus checking that the consignment has travelled to the proper disposal site.

These 'trip-ticket' regulations are concerned with a minority of hazardous wastes, as approximately 60 per cent of hazardous wastes are 'non-special', and their movements are therefore not recorded through the consignment note system (see Figure 4.1). The boundary between wastes being relatively inert or hazardous is fuzzy as there is an important circumstantial element – most waste streams have the potential to do harm to the environment if deposited in the wrong place at the wrong time. The uncertainty is also increased by the lack of reliable information on the total tonnages of industrial and commercial waste (see Figure 4.1) of which hazardous wastes are a subset (Royal Commission on Environmental Pollution 1985: 27). Without such information the Royal Commission did not consider it possible for waste disposal authorities to properly prepare waste disposal plans for their areas. They did not comment on the way in which data from the Special Waste Regulations were applied to waste disposal plans.

The Special Waste Regulations rely heavily on the obligations on producers, transporters and disposers to pass on accurate information to the authorities. The users of the system thus supply the initial information required to bring them within the system. The Joint

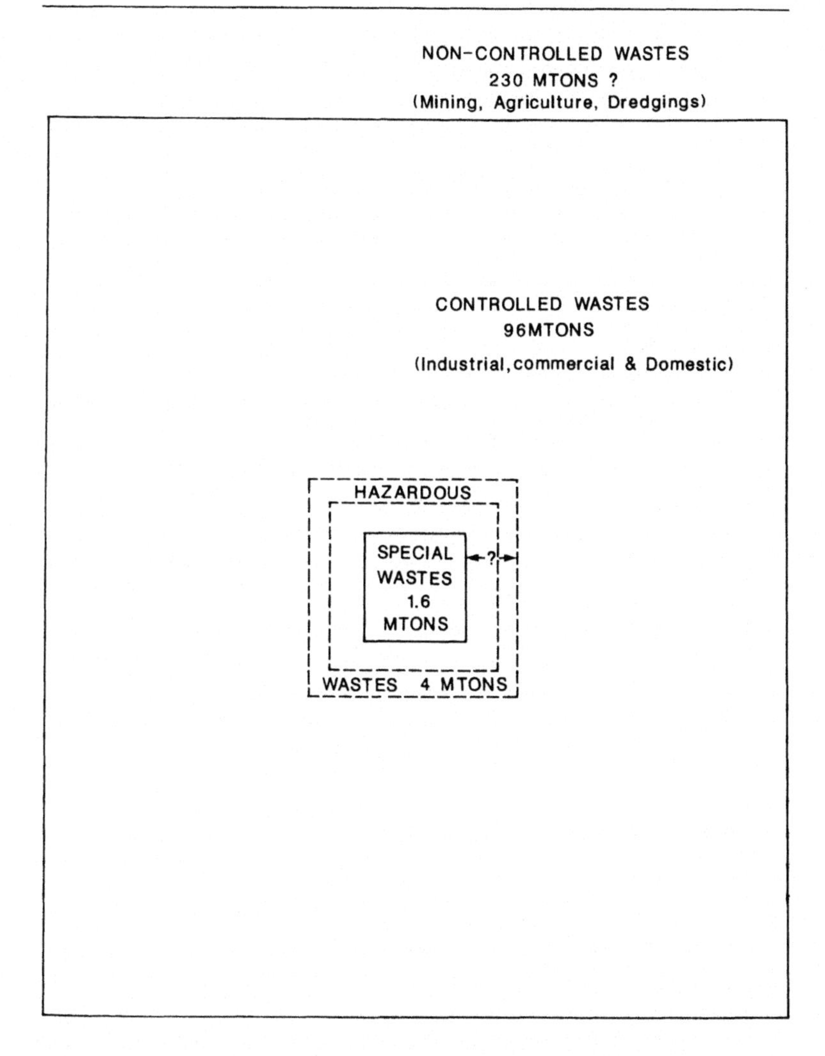

Figure 4.1 Estimated annual waste arisings (Royal Commission on Environmental Pollution, 11th Report, 1985)

Review Committee of the Special Waste Regulations commented that:

> no such system of special control could effectively operate if authorities were confined to being passive recipients of notifications

> under the Regulations and were not able to take the initiative in seeking other relevant information or make enquiries designed to verify information already received.
>
> (Department of the Environment 1985: Annex 8/5)

A significant restriction on the activities of waste regulation is that waste regulators and waste planners cannot gain access to names and addresses of firms from Census of Employment data held by the Department of Employment. Waste planning and waste intelligence gathering activities are not given equal status in law to other planning activities. The importance of this point relates particularly to waste regulation in large metropolitan areas.

London may contain as many as 170,000 companies of which approximately 1,000 are known special waste producers. Certain industrial activities are directly associated with the production of special wastes (for example, metal finishing, pharmaceutical manufacture, research and development laboratories) and locating all firms of a particular sector in order to assess their waste streams is made more difficult without full access to official statistics and addresses. This is a case of one set of legislation (Control of Pollution Act 1974 and subsequent regulations) being undermined by another (Statistics of Trade Act 1947, Employment and Training Act 1973, Employment Act 1988). It is also indicative of the low priority assigned to waste planning in the United Kingdom.

WASTE PLANNING AND REGULATION IN THE GREATER LONDON AREA

The London area is one of the largest special waste producing regions of the country, and currently 94 per cent of the special wastes arisings are exported outside the area (the majority deposited within being asbestos). When the DPWA was first implemented (1973) a third of the waste classified as notifiable under that Act was deposited within the Greater London area. The statistics suggest a very rapid decline to 14 per cent in 1974 and 6 per cent the following year. The number of licensed landfill sites within the London area declined from over seventy in the early 1970s to currently about twenty. The overall trend is for waste to be transported further afield as the sites closest to London are used up.

The dwindling supply of nearby landfill is an important factor in the disposal costs of the high volume wastes – such as construction and demolition wastes. Although the proper disposal of hazardous wastes is more complex than the considerations for inert categories of wastes, the availability of proximate sites for treatment or disposal may also be a problem: not least through safety considerations. A hypothetical situation where the most toxic wastes might travel the furthest distances from source to sink would increase the probability of accidental human exposures during transportation. Ideally these risks should be balanced against the benefits of pursuing a particular disposal or treatment option in terms of the 'best practicable environmental option' (Royal Commission on Environmental Pollution 1985: 41). This strategic problem was considered in the Greater London Council's consultation document 'No Time To Waste' in which it was proposed that a register of hazardous waste producers and handlers should be kept (no such register has been created):

> Most hazardous wastes from London are taken for deposit by landfill in the Home Counties. Relatively small quantities are treated in chemical treatment plants or incinerators almost all of which are large distances from London. One of the purposes to which a hazardous waste register would be put would be to determine the need for the provision of a strategic hazardous waste treatment plant and other disposal facilities in London.
>
> (GLC 1983: 6.28)

At abolition of the Greater London Council in 1986 the integrated waste disposal system for London was divided between sixteen new WDA's comprising individual London Boroughs (arranged into four voluntary groups) and four statutory groups. Overall regulation remained with a single body – the London Waste Regulation Authority (LWRA). The new authority took over the GLC's Hazardous Waste Unit and has continued to operate the same computer system for consignment note record-keeping. The strategic planning function for London was not retained by the LWRA, and responsibility for waste disposal plans was devolved to the four voluntary and four statutory groups.

The new arrangements for waste disposal in the former Metropolitan Counties were reviewed in April 1989 by Her Majesty's Inspectorate of Pollution. Nearly all statutory and voluntary groups in London claimed to have waste plans 'under preparation' in order to meet the Department of the Environment's deadline for unfinished plans (October 1989). Only two of the planning areas in London actually met this date, and for four areas no survey work had been started by September 1989. The Inspectorate recommended a return to strategic waste planning for Greater London in its report:

> Waste management in London is highly complex involving a web of movements generally to remote sites. In the Inspectorate's opinion a strategic view is needed and accordingly recommend that the LWRA prepares an overview waste disposal plan for the London area.
>
> (HMIP 1989: 55)

The analysis of special waste arisings for the Greater London area provides insight into what the benefits might be of collecting data on a regional basis for the purposes of integrated waste planning. As outlined above, there is at present no integrated planning system for waste disposal in London to utilize such data. In the analysis of London data here, total tonnages of wastes regulated since 1973 are briefly discussed, followed by the detailed analysis of waste compositions and destinations for the year 1983.

ANALYSIS OF HAZARDOUS WASTE STATISTICS FOR LONDON

The use of data from either the Deposit of Poisonous Wastes Act or the Special Waste Regulations in most waste disposal plans has been limited to the reporting of total tonnages of hazardous wastes (for example, Humberside Waste Disposal Plan 1982: 15). Both pieces of legislation have generated immense quantities of detailed information on the nature and disposal of hazardous wastes, yet neither have resulted in an improvement in knowledge of waste production beyond the reporting of annual total tonnages for notifiable or special waste arisings.

The passing of the DPWA represented a loss in potential knowledge, as the universe of waste under its control was far greater than that of the Special Waste Regulations (SWRs). The learning potential of the DPWA was not realized, as the notifications received by WDAs were never analysed comprehensively nor was a common notification format imposed across the country. There was no body of established knowledge of hazardous waste arisings and disposals from which the SWRs could be drawn up (Finnecy 1985: 7). The localized nature of many WDAs, the miscalculation of the size of the hazardous waste regulatory problem (WDAs could not cope with the paperwork) and the *laissez-faire* policy of Central Government all contributed to this failure.

TREND DATA WITHIN WASTE CLASSIFICATION FOR THE DEPOSIT OF POISONOUS WASTES ACT AND THE SPECIAL WASTE REGULATIONS

The only data regularly computed from the DPWA or the SWRs by either the GLC or the LWRA have been the annual tonnages of wastes arising, imported, exported or deposited within Greater London, broken down by a system of waste classification. These data were produced for the DPWA from 1973 to 1980 (using a card system and hand counts) and for the SWRs from 1981 onwards (using a fairly sophisticated but slow and inflexible computer system).

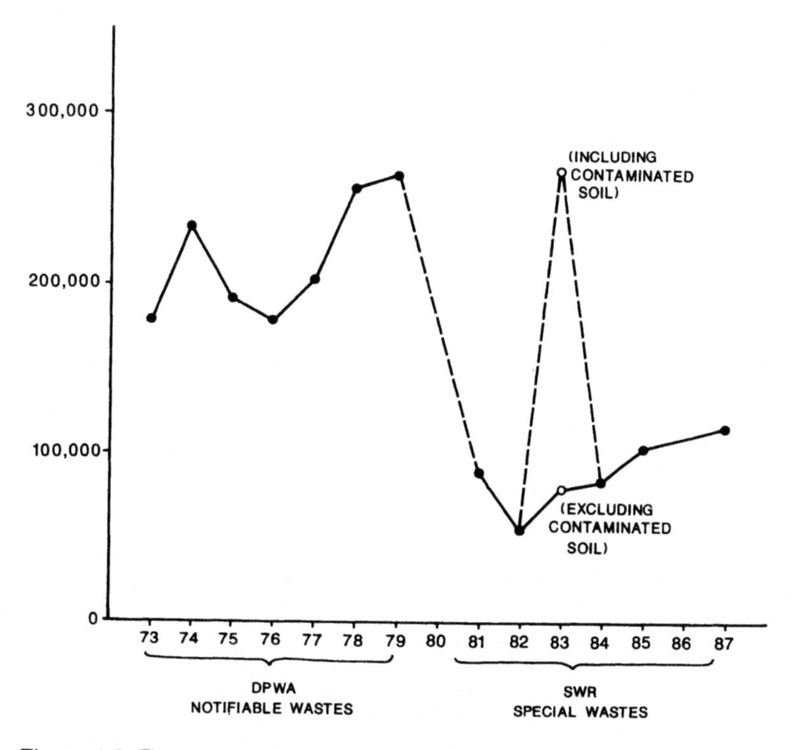

Figure 4.2 The total tonnages of hazardous waste regulated under the Deposit of Poisonous Wastes Act and the Special Waste Regulations arising from London

Although these tonnages have been computed, they have not been analysed by either the GLC or the LWRA for long-term trends. Since 1973 the data have been reported quarterly with comparative data from the previous year only. Using these records the overall trend in hazardous wastes arising from within Greater London was established. This was found to be upward – as indicated by tonnages notified under the DPWA (1973–80) and the SWRs (1981–7) in Figure 4.2. This trend appears to have been little affected by the two major economic recessions since 1973.

The switch from the DPWA to SWRs in 1981 led to a 60 per cent reduction in the tonnage of waste consignments being directly reported to the GLC by waste producers (see Figure 4.2). This picture is complicated by the total tonnages for 1983, when a consignment of nearly 200,000 tons of contaminated soil was being removed from the site of a disused gas works at Wandsworth.

There is a qualitative dimension to the hazardous waste regulation – 10 tons of unbroken asbestos sheeting does not represent the

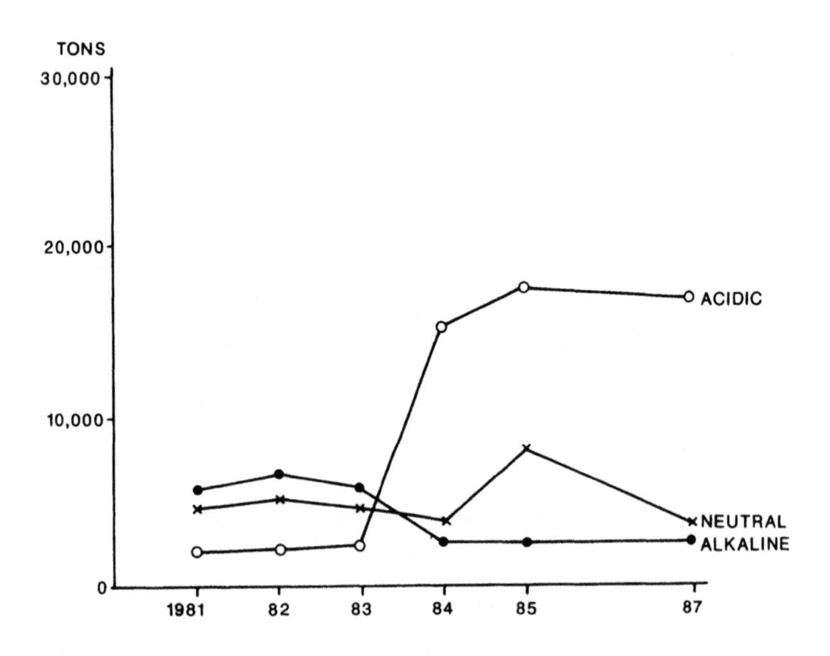

Figure 4.3 The total tonnages of aqueous special wastes arising from London

same degree of hazard to either health or environment as 10 tons of acidic liquid wastes contaminated with heavy metals. It is therefore surprising that the reporting of special waste data, either nationally or regionally, has not focused on the trends within particular categories of waste. Consignment note data for aqueous toxic wastes are presented here as an example of regional trend analysis by composition.

The total tonnage of aqueous special wastes arising from London has increased rapidly since 1983 (see Figure 4.3). The most significant contribution has been from acidic waste streams (pH <4), with neutral ($4 < \text{pH} < 10$) and alkaline waste streams pH >10) remaining fairly level from 1981 to 1987. The balance between acidic waste streams and alkaline is important nationally, as the two can be used to neutralize each other either on site or at a chemical treatment works. Typically the volume of alkaline waste streams produced is slightly below that of acidic waste streams. This is somewhat contradicted by the tonnages reported under the DPWA, which were much greater for alkaline waste streams than acidic from 1977 onwards. The tonnages of acidic special waste streams have now risen to a level which is comparable to the maximum reported under the DPWA.

The main treatment and disposal options applied to acidic waste streams arising from London in 1983 were chemical treatment (62 per cent of waste streams) and direct co-disposal with household wastes in landfill sites (25 per cent). The majority of London's aqueous acidic special wastes (72 per cent of waste streams) went to a few sites in Essex, and those waste streams that were directly co-disposed would have been sent to two particular landfill sites.

The analysis of consignment notes cannot reveal whether or not waste streams were diluted before disposal, or neutralized with lime, as that would have been a function of on-site management. However, increased acidity in a landfill site practising co-disposal can increase the mobility of certain heavy metals, such as cadmium, zinc and nickel, which reduce the attenuative properties of the landfill. This effect is a function of the concentration of the acids, the loading rate and whether or not acids have been mixed (a mixture is generally more reactive). Other hazards associated with landfilling of acids include local toxicity to microbiological populations, and the production of toxic gases after reaction with certain other wastes, such as those containing strong reducing agents and halogenated organic compounds (Department of the Environment 1986: 107).

The environmental significance of the increased tonnage reported from consignments of acidic wastes cannot be investigated here. The upward trend in arisings was reported to the London Waste Regulation Authority and Thames Water. Neither authority were aware of the trend. The most plausible explanation is that the Thames Water Authority might have made changes to discharge consent conditions, leading to fewer acidic waste streams being discharged to drain and more being taken off-site in tankers. If this is true, it raises important questions about the partitioning of pollution control between media, when such pronounced inter-media transfers might be occurring undetected.

A further possibility would be that this is not a real effect, but has resulted from an increase in the regulation of acidic waste streams that were already being produced, or from a change in reporting procedure or waste classification, or both. Further investigation is required, but whatever the most appropriate explanation, the example illustrates the importance of carrying out such statistical analysis on a regional basis.

THE DETAILED ANALYSIS OF CONSIGNMENT NOTES ARISING FROM GREATER LONDON IN 1983

Official hazardous waste statistics focus almost exclusively on the presentation of tonnage data (Hazardous Waste Inspectorate 1988: 7). This can hide important features of the problem of waste regulation. Often it is relevant to count the number of different waste producers or discrete waste streams contributing to a particular category of special waste. For instance, small, infrequent consignments of highly toxic wastes from numerous and diverse producers represents a regulatory nightmare, compared with large tonnages from waste streams arising at regular intervals from a smaller number of conspicuous producers. The units of analysis used here are tonnages, number of waste streams involved and number of premises. Adopting this mixed approach is an attempt to learn more about special waste production from this area than official tonnage statistics reveal.

The total tonnage of wastes covered by the analysis represents only 13 per cent of the official statistic for 1983 (268,657). The difference is mostly accounted for by a large one-off consignment of contaminated soil which was not reported in this particular computer file, and therefore does not form part of this analysis (this

Table 4.1 Special waste consignment note analysis: summary statistics

Total tonnage in analysis:	34,800
Total number of waste streams:	2,475
Total number of premises:	970

originated from Wandsworth Gas Works – a site which was redeveloped by the GLC into a modern riverside transfer station). The data represents the majority of waste streams (2,475 streams) and waste producers (970 producers) for this particular year (see Table 4.1).

Aqueous wastes (acids, alkaline and neutral liquids) were the largest group of special wastes arising by weight in 1983 (44 per cent of total tonnage), and solid wastes accounted for most of the rest (40 per cent) as shown in Table 4.2. Asbestos was the main constituent of the latter (93 per cent of total solid tonnage).

Analysis by number of waste streams within each grouping revealed differences between liquid toxic wastes and solvents. The former accounted for the largest proportion of the total tonnage (44 per cent), and the share of the number of streams was 25 per cent of that total. In contrast, 21 per cent of waste streams contained solvents, yet only contributed 9 per cent of the total tonnage. The mean weight per waste stream was 35 tons for liquids and 10 tons for solvents (see Table 4.2). Solvents were therefore more often found in smaller volume waste streams than was the case for liquid special

Table 4.2 Special waste consignment note analysis: analysis of waste compositions

Physical characteristic/ composition	*% Tonnage/ year*	*% Waste streams*	*Mean tons/ waste stream/year*
Liquid – aqueous	44	25	35 tons
Solid	40	43	11 tons
Solvents and solvent sludges	9	21	10 tons
Oils and oily sludges	2	3	40 tons
Contaminated industrial e.g. contaminated soil	2	4	8 tons

wastes, as the latter were more likely to be bulk wastes related to production processes. This was reiterated by the mean tonnages per site per year. The mean for the 182 premises contributing solvent-bearing special wastes was 13 tons/producer/year, and for the 168 producers consigning liquid special wastes: 130 tons/producer/year.

These two groups represent different poles of the Special Waste Regulation universe – the large-scale production process wastes, which mainly arise as a consequence of manufacturing processes on site (for example, inorganic acidic wastes from a metal-finishing process), and ubiquitous wastes which often arise coincidentally to the main business of an establishment (for example, machine-degreasing solvents). The latter group in particular represents a challenge to the ingenuity of the waste regulator, as non-point sources of arising are more difficult to identify and bring 'within the net'.

The sum total of a large number of non-point sources evading control could be of greater significance to environmental protection objectives, such as the prevention of ground-water pollution by micro-pollutants, than the contributions of conspicuous institutions, who might be more inclined to observe the rules to escape public criticism. The regulation of non-point sources of special wastes under the voluntaristic workings of the Special Waste Regulations is particularly difficult in metropolitan areas. Such areas are likely to contain high concentrations of small firms unaware of the regulations. Estimation of the number of firms producing hazardous wastes requires regular waste producer surveys to be conducted, but few waste regulation authorities have either the expertise or resources to carry out such work.

Destination of special wastes

Analysis of consignment note records for 1983 revealed that half the tonnage of special wastes was being exported to Essex and 25 per cent to Kent (see Table 4.3). The North of England and Scotland were receiving about 4 per cent. In terms of strategic waste planning, the disposal and treatment of special wastes is a regional issue.

The London area has become heavily dependent on its neighbours for disposal outlets, particularly on Essex. The use of Essex as a destination for London's wastes has developed over the centuries. As London's population grew it became unacceptable to throw all solid wastes into the Thames and instead some wastes were taken

Table 4.3 Special waste consignment note analysis: location of sites handling London's special wastes in 1983

Location	*% Total annual tonnage*	*% Total waste streams*	*Mean tons/ waste stream/year*
Essex	50	34	21 tons
Kent	25	9	39 tons
Other SE	11	33	5 tons
North & Scotland	4	5	10 tons
London	6	16	5 tons
Midlands & Wales	1	4	2 tons

Note: These statistics do not necessarily represent the location of ultimate disposal, as some sites are transfer stations or treatment works.

down stream by barge to be dumped on the Essex Marshes. The dominance of Essex sites as destinations for toxic waste disposal partly relates to this historical aspect, but also to the underlying geology. The clays in the area are suitable for attenuate and disperse landfill sites. The rate of migration through these clays is sufficiently slow to allow attenuation of many of the hazardous components of waste leachate to occur.

Disposal and treatment methods

Table 4.4 shows that landfill was the main disposal route for special wastes in 1983 (63 per cent of total tonnage, 41 per cent of waste streams). Lagooning of special wastes was a relatively important method in tonnage terms (19 per cent), but involved a smaller proportion of waste streams (8 per cent). Conversely, incineration and chemical treatment accounted for significant proportions of the total number of waste streams (25 per cent and 20 per cent respectively), but represented proportionately lower total tonnages (8 per cent and 6 per cent).

These findings reflect the relative costs of the major disposal routes, with the options which account for the greatest total tonnages being the cheapest. Landfilling costs for toxic wastes fall within the range £2.50–£35/tonne, incineration between £40 and £900, and chemical treatment from £10 to £390 (Wynne 1987: 201, all costs exclude transport).

Table 4.4 Special waste consignment note analysis: disposal/ treatment methods for London's special wastes in 1983

Method	*% Tonnage*	*% Total waste streams*	*Mean tons/ waste stream/year*
Landfill	63	41	24 tons
Incineration	8	25	5 tons
Chemical treatment	6	20	4 tons
Lagooning	19	8	66 tons
Fixation	4	4	13 tons

The analysis of disposal and treatment methods by the location of sites handling those wastes has been based on the number of waste streams involved (see Table 4.5). Some options required wastes to travel long distances. This is particularly noticeable for incineration, where nearly one-fifth of waste streams analysed travelled outside of the south-east of England in 1983 (the main destinations were Wales and Scotland). Some of the most toxic waste streams, those that are incinerated and not 'co-disposed' in landfill sites, appear to have travelled the furthest from London for disposal due to the limited incineration capacity within the region. This capacity has been reduced further since 1983, with the closure of an incinerator in the London Borough of Havering which was licensed to burn certain categories of hazardous wastes.

Table 4.5 Special waste consignment note analysis: disposal and treatment methods by location
Base: 2,400 waste streams

	% of waste streams handled in:				
Method	*Essex*	*London*	*Other SE*	*North*	*Midlands/ Wales*
Landfill	35	17	44	2	1
Incineration	11	28	42	7	12
Chemical treatment	46	1	46	1	6
Lagooning	95	—	5	—	—
Fixation	83	1	—	—	8

The primary importance of Essex to bulk-liquid special waste disposal can be inferred from the high proportion of waste streams arriving there for lagooning and chemical treatment. One landfill site at Pitsea takes the majority of these wastes, and in 1983 this amounted to about 6,000 tons (equivalent to 40 per cent of the liquid wastes in this analysis). There is probably an imbalance between acidic and alkaline liquid wastes being received at this site, as suggested by the trend analysis (see Figure 4.3).

The dependence on a few large sites within one county for a high proportion of disposal or treatment of certain categories of special wastes may cause problems in the long term, particularly as Essex County Council is opposed to the future location of any similar sites in the County:

> Essex should in the immediate future (i.e. for the next five years) continue to accept the disposal of hazardous waste (including high hazard waste) brought into the County from London, Hertfordshire, Cambridgeshire ... on the understanding that these areas seek to be self-sufficient within the period Essex has agreed to accept (i.e. five years) ... With the exception of London and the named neighbouring counties, the import of waste from other areas of the country shall cease as soon as possible.
>
> (Extract from Essex C.C. waste disposal plan – July 1979, submitted as oral evidence to the House of Lords Select Committee on Science and Technology 1981)

Greater London will continue to deposit the majority of its hazardous wastes outside of its boundaries as there is insufficient space for more landfill sites. It is possible that within the next decade disposal and treatment costs will have to rise steeply unless new options are developed within, or adjacent to the area.

CONCLUSIONS

The consignment note system of the Special Waste Regulations, although imperfect, has been under-used as a data source on hazardous wastes. Two factors undermine the comprehensiveness of the data source: special wastes do not include environmental hazards, and there are likely to be significant numbers of non-regulated special waste producers, due to the potentially large number of small producers ignorant of the complex regulations.

Despite these inadequacies, the analysis has detected important trends in tonnages arising from the London area, and has described the regional dimension of consignment movements. On the one hand, the region is highly dependent on a few sites in Essex licensed to accept hazardous wastes. On the other hand, wastes for incineration travel great distances outside of London's immediate surrounds due to the lack of high temperature incinerators within the region.

The current arrangements for waste planning in London are officially recognized as being inappropriate to the complexity of the situation, and should be carried out on a regional scale. The intelligence gathering and planning functions of waste regulation must be afforded greater priority in both resourcing and legal standing. Without these necessary changes, there is a danger that new legislation and future strategic developments in waste treatment and disposal will founder through lack of solid information.

REFERENCES

Department of the Environment (1985) *Control Over Dangerous and Other Harmful Wastes*, Report of a Review of the Control of Pollution (Special Waste) Regulations 1980, London: HMSO.

Department of the Environment (1986) 'Landfilling Wastes', Waste Management Paper no. 26, London: HMSO.

Finnecy, E.E. (1985) *Pollution by Wastes*, Harwell: United Kingdom Atomic Energy Authority.

Greater London Council (GLC) (1983) *No Time To Waste, A Limited Planning Statement for Waste Regulation, Recovery and Disposal 1983–2003*, London: GLC.

Hazardous Waste Inspectorate (1986) *Hazardous Waste Management: 'Ramshackle and Antediluvian'?*, Second Report of the HWI, London: DoE.

Hazardous Waste Inspectorate (1988) *Third Report*, London: HMSO.

Her Majesty's Inspectorate of Pollution (HMIP) (1989) 'Waste disposal regulation and operation in the former Metropolitan counties: a review'. London: Department of the Environment.

Humberside Waste Disposal Plan (1982) Humberside County Council.

Royal Commission on Environmental Pollution (1985) *Eleventh Report, Managing Waste: The Duty of Care*, London: HMSO.

Wynne, B. (1987) *Risk Management and Hazardous Waste*, Berlin: Springer-Verlag.

Chapter 5

The development and application of geographical information systems in waste collection and disposal

Linda Crichton

INTRODUCTION

Much of the data relating to the wastes management industry are inaccurate or poorly quantified and only recently has more reliable data on the basic aspects of the industry such as its size, distribution and value, and also the nature, range and quantities of material handled emerged. This is an important characteristic of the industry, and one that sets it apart from most other industrial or commercial activities.

Locating information relating to waste generation and disposal in the UK can present considerable problems. Data tend to be widely dispersed, and there is a paucity of published material. Many of the data available are incompatible, inconsistent or related to a limited period in time. A number of special studies have been undertaken to review aspects of wastes management including the Select Committee of the House of Lords (the Gregson Committee), the Royal Commission on Environmental Pollution study, the Audit Commission, and the recent House of Commons enquiry into hazardous wastes. Other information of a similar nature can be obtained from the annual reports produced by the Hazardous Waste Inspectorate (now incorporated in Her Majesty's Inspectorate of Pollution, HMIP).

The largest body of published information on household and commercial waste collection and disposal was the Chartered Institute of Public Finance and Accountancy (CIPFA) statistics which were published on an annual basis for England and Wales. Similar data are not available for Scotland or Northern Ireland (however, financial data are published). They provided a broadly accurate indication of household wastes handled in the public and private sectors. However, although the data is still collated, it is no longer publicly available for reasons of commercial sensitivity arising from the compulsory competitive tendering of local authority services. However, with only just over 40 per cent of household wastes actually weighed, many of the returns are estimates based on the payload of the vehicles and the number of trips to the disposal facility.

In contrast, the overall quality of information regarding commercial and industry waste arisings and disposal is extremely poor. Few waste producers weigh their wastes or have the facilities for doing so, and the same is true of most contractors and most sites accepting wastes for disposal. Site operators may be required, under the conditions of the site licence, to make returns to the waste disposal authority (WDA) – the licensing authority – of wastes deposited. However, the figures supplied take a wide variety of formats, and are of variable accuracy.

These problems, however, have not gone unnoticed. In particular, the Eleventh Report of the Royal Commission on Environment Pollution focused attention on the inadequacies and inaccuracies of information available on the quantities of controlled wastes being produced in the UK, and expressed the need for a coherent scheme for the recording of information on arisings, movements and disposal of controlled municipal wastes,[1] and at an early stage in its development the Hazardous Waste Inspectorate identified as a priority task the creation of a comprehensive hazardous waste management database. As the industry has developed, the need for more accurate data has increased, both in the public and private sectors, to satisfy a number of requirements:

- To provide better management information
- To facilitate forward planning
- To improve the accuracy of costs and pricing structures
- To maximize the use of available disposal facilities

This paper describes the work undertaken by Aspinwall & Company, an environmental management consultancy, to improve the quality of information in two areas of wastes management which have been recognized as requiring detailed attention. The first is household waste generation, in particular research to understand better the quantity and quality of wastes being generated by individual households. The second is the creation of a comprehensive waste management information system.

HOUSEHOLD WASTES

Since 1983 Aspinwall & Company, as part of the Department of Environment's programme of waste management research, has been making detailed investigations into household waste generation. The principal aims of this work have been to obtain more accurate information on the quantities of wastes being produced by British households and a better understanding of the variations in output between households in different socio-economic groups and in different regions of the country.

The approach developed to tackle this problem uses some standard methods and makes use of commercially available computerized information sources: the Post Office's postcode information system and the ACORN system of neighbourhood classification developed by CACI Ltd. In the early stages of this research certain principles were developed which have formed the basis of all subsequent work. It is not the purpose of this paper to describe the development of these principles – that has been well documented elsewhere.[2,3] The principle features are, however, noted below:

(i) The individual household is taken as the basic unit in generating wastes. The amount of waste produced by a household is measured in kilogrammes per household per week (kg/hhd/wk).
(ii) Census data is used as it is the most uniform and comprehensive national source of information collected at household level.
(iii) The ACORN system is used as it provides a means for classifying any residential area in the country on the basis of its socio-economic characteristics and it has been proved to be powerful discriminator of consumer habits and the purchasing power of households.[4,5]
(iv) The Post Office's postcode address file is used to obtain an accurate household count for any defined area in the country.

This information provides us with an accurate household count for a local authority area and details of its socio-economic characteristics (given by the numbers of households represented by each ACORN group); the final information requirement is accurate weight data. For those authorities with access to a weighbridge the weight of a load deposited by every vehicle entering a site can be determined and aggregated over any time period.

Local authorities arrange their refuse collection services so that every vehicle operated by the authority has a set pattern of work for each day of the week, and this schedule is repeated on a weekly basis. So just in the same way as the ACORN profile and household count for the authority as a whole is calculated, the same information is aggregated at a far smaller scale, that of the route of each collection vehicle on a daily basis. In a typical urban area this may be in the order of 800 to 1,000 households per day depending on the method of collection. However, the number of ACORN groups represented will be much fewer than for the entire area and in some cases only one ACORN group may be represented.

This information can then be matched with the weight information recorded for each vehicle on a particular day. A statistical model has been developed to enable household waste factors to be computed for each ACORN group represented in the authority. Typically a full year's data is required to enable any seasonal variations in waste output to be detected and the factor calculated will represent the average weight of waste generated by a typical household on a weekly basis.

Over twenty local authorities in Britain have co-operated in this research. These authorities have ranged in size and socio-economic characteristics from a London Borough to a small rural district in Wales to a large Metropolitan Borough in Lancashire. They also operated a number of different collection systems (for example, wheeled bins, plastic sacks, conventional dustbins) and collection policies (for example, collections restricted to one container per household, no garden waste collected). The results, in the form of ACORN waste factors, have shown conclusively that household waste generation does vary according to the socio-economic characteristics of the households and with more affluent households typically producing more waste than less-well-off households.[6,7] For example, a 50 per cent variation in household waste output was measured in Bury Metropolitan Borough Council[8] which operates wheeled bin collections. The degree of variation between households, however, is related to the method of collection employed by the local authority and greater variations tend to occur with wheeled bins. Compositional analyses of wastes (carried out by Warren Spring Laboratory) from household samples selected to represent different ACORN groups have also revealed variations in waste composition between 'better-off' and 'less-well-off' households.

Refuse collection planning and management system

These investigations and findings together with the ability to characterize existing collection rounds opened up the possibility for detailed refuse-collection round planning using the ACORN/postcode data as the building blocks. The combined ACORN/postcode database supplies the following information at postcode level:

- Street name
- Property name or number range
- Household count
- Postcode
- ACORN classification

The data can be manipulated into postcode, refuse-collection round and alphabetical street order. Further fields have been added to the database to allow records to be assigned to a refuse-collection round and to enable local authorities to enter data relating to standard minute values and other performance criteria. Routines exist for entering and analysing information on trade premises. Waste factors for each ACORN group represented within an authority can be entered into the system and used to calculate the estimated waste output both for the authority as a whole and on a round-by-round basis. This provides useful management information for improving the efficiency of the vehicle fleet by planning for each vehicle to maximize its payload.

Given this database it is possible to begin to build up rounds by specifying in the search specification the area to be searched as defined by postcodes and criteria such as:

- Number of households
- Time constraints
- Payload capacity of the vehicle
- Need to maintain geographical continuity

These routines allow the user to try and test various options and determine the most efficient collection service in terms of vehicle use and manpower. The computerization of the basic information required for refuse collection planning offers the flexibility to manipulate the data easily and effectively and also enables the schedules, once completed, to be maintained and updated as and when necessary.

The system is marketed under the name of ULTRA and is being

used successfully by a number of local authorities. The City of Nottingham has introduced wheeled-bin collections across the whole of the city, some 120,000 households, and has used the system to reschedule the entire collection service, the average time to plan a round having been reduced from five days to two days. Other authorities have used the system to help put their house in order prior to going out to competitive tender and also to help in the preparation of tender specifications and documents.

WASTE INFORMATION DATABASE

An important application of computer technology to the area of data and information management in the waste industry is SITEFILE, a computerized database of all licensed waste disposal and treatment facilities in Great Britain. The database has been assembled from information gathered from the public register of waste disposal licences held by every WDA in the country – the county councils and metropolitan districts/boroughs in England and the district councils in Scotland and Wales. The information was first collected during 1984 in order to supply the then Hazardous Waste Inspectorate with a listing of all waste disposal facilities in the country as such data was not readily available.

This exercise uncovered a tremendous variation in site licences and gathered so many waste descriptions that within the time available before publication it was impossible to produce a meaningful system of classification to compress the data into a register of manageable proportions. This, together with the problems and costs associated with assessing each individual register every time the information was required, inhibited much more than a county-only search.

SITEFILE overcame these problems by providing a logical and common format to record and reproduce information from each site licence with the minimum of interpretation. There are around 600 waste classes and individual waste descriptions, of which over 30,000 can be identified separately if required. Currently SITEFILE contains 9,500 records of which over 5,000 represent operational facilities. The remaining records are for closed sites, sites not yet started or sites where the licence has been surrendered or revoked. Fifty types of facilities can be identified including landfills, incinerators and pulverizing plant, and storage and treatment plant.

The main feature of SITEFILE is the ability to search the database, select and present information. Basic search routines include the ability to search by area (defined by boundary of the licensing authority), operator, type of facility and authorized wastes. Additionally, the specification may eliminate from the search sites for the disposal of in-house produced waste only, closed sites or sites which prohibit specified wastes; nominated disposal contractors, and private sector or public sector sites only, can also be searched for.

The record of sites which match the specification are printed in county order using a standard print format which contains the national grid reference (NGR), details of the licensed operator and contact address, a comprehensive waste profile (authorization, waste description, specified physical form and quantity), the site category and other licence particulars. The advantage of these profiles is that they present information from the very varied licences in a consistent, easy-to-follow and comparable format.

More complex procedures are available to analyse information within any size of area without being restricted to the geographical area defined by the administrative boundaries. Searches can be performed about a point determined by a standard eight digit NGR. Using this method it is possible, for example, to study disposal patterns in relation to the points of waste generation or the distribution of types of sites or operators within an area. All the sites found in the search can then be displayed graphically on computer-generated maps. A hypothetical search is reproduced below:

Example

Assume a quantity of slag is produced in Northfleet in Kent and requires to be disposed of as near to the point of production as possible. The grid reference of the point of production is TQ 650 740 and a search is performed around a 10 km radius of this point.

Results

The search produces a total of fifty-three sites; however, only twenty-seven are in Kent with the remaining sites being located across the Thames Estuary in Essex. If we consider the Kent sites only and eliminate the closed sites we are left with twenty of which seventeen are landfills.

> A further refinement of the search is required to find sites which are able to accept slag or under the Kent classification of waste types, Kent Category B. This produces seven sites of which one accepts in-house produced waste only and two are dormant as of December 1988. This leaves four available sites for the waste contractor to approach.

A well-designed and flexible information system readily enables add-on facilities. Natural extensions to the system have been developed and each in response to user requirements. These include modules for report or inspection routines allowing users to record site-specific information against a particular date; waste input records enabling the recording of input data for a site (weights, volumes, prices, position of deposits); a module to record the details of transfrontier shipment notes for imported wastes, and a similar routine for the special waste consignment system; and finally a module to enter details of locations that are not licensed but which may be of interest to the user, for example, old sites which may be generating landfill gas, possible future sites and illegal tipping areas.

COST EFFECTIVENESS

When used efficiently, information systems/databases are widely recognized as being cost effective. With those remarks made in the introduction to this paper in mind, their use in a cost competitive industry which is bound to respond to environmental pressures placed upon it is readily apparent.

With respect to the system for refuse collection, cost savings can be realized in two areas. First, administrative costs incurred in the planning and management of the service. The fact that an authority can purchase a database for its entire area together with the software to manipulate it means that work can commence immediately on planning or rescheduling rounds and the time and costs involved in collecting detailed survey information are eliminated. With the current pressures on both public sector cleansing services and the private waste contractors to respond to the legislation regarding compulsory competitive tendering of local authority cleansing services, the need for sufficient and accurate information within a short period of time is more important than ever.

The second area for cost savings is by improving the efficiency of the service. The majority of authorities purchasing the system are doing so because they are about to make changes, either through implementing a new collection system, or undertaking a rescheduling exercise to increase the efficiency of the service. If by using the system the number of collection rounds is reduced by one this can represent a saving of upwards of £70,000 per annum – the cost of operating one refuse-collection vehicle.

Implications for environmental management, however, are most directly realized through the use of these techniques for the planning of wastes disposal and management. Information on the quantities and composition of household wastes is essential where the wastes form the feedstock to mechanical plant, especially where the requirement is to recover fuel and or materials to specifications conforming to market requirements. Until recently the design and site selection of plant such as refuse derived fuel (RDF), transfer stations and materials reclamation plant have not necessarily had the benefit of reliable information on the quantities and composition of wastes to be accepted.

SITEFILE has also proved to be a cost effective tool for a number of users as before the existence of the database someone requiring information on disposal facilities at more than the local scale would have been involved in numerous hours of research and many phone calls. Current users tend to fall into four main categories:

(i) Waste contractors, disposers and hauliers for whom SITEFILE can provide a new perspective to market analyses and business planning as a regional and national picture of the disposal market and the key parties involved can now be readily gained.
(ii) Waste producers who often lack direct information on disposal outlets for their wastes but who now have to accept greater responsibility for the disposal of those wastes under the Environmental Protection Act 1990.
(iii) Public authorities and institutions including Central Government and the WDAs whose requirements are many and varied. It is increasingly an important tool for bodies which own large land banks and whose area crosses into several administrative units. Landowners or property developers with national, regional or local interests can have established for them the proximity of waste facilities. Recent interest has focused on landowners with property which might suffer the particular effects of landfill gas migration.
(iv) A further recent use of the system is as a marketing tool for

companies selling products and services to the waste disposal industry – monitoring equipment, liners for landfill sites, landfill gas services and so on.

The fact that these systems are successfully marketed to both the public and private sectors, and have applications not conceived of at their development stage, is proof alone of the cost effectiveness and value of these GIS/databases as applied to wastes management.

CONCLUSION

Geographical information systems are beginning to play a major role in the management of our environment through the cost effective planning of wastes management. The availability of improved information provided in this way allows for the fuller consideration of the options for wastes management towards an improved environment. The further use of such tools must be encouraged by all those who seek to influence and improve the environment of tomorrow. Increasing environmental controls such as those for discharges into water and emissions into the atmosphere will further highlight the need for information systems on which to model likely future conditions.

NOTES

1 Royal Commission on Environmental Pollution (1985) *Eleventh Report, Managing Waste: The Duty of Care*, London: HMSO.
2 Davies, D.R. (1983) 'The use of census data in household waste prediction', Department of the Environment and Institute of Wastes Management Joint Seminar on Progress in the Forecasting and Analysis of Household Waste, Aston University, September, 1983.
3 Davies, D.R. and Marsh, R.J. (1986) 'Progress in wastes analysis and the forecasting of UK waste arising', Paper presented to the Institute of Wastes Management Annual Conference, Bournemouth, 10–13 June 1986.
4 CACI Market Analysis Division (1988) *ACORN: A New Approach to Market Analysis*, London: CACI Ltd.
5 CACI Market Analysis Division (1988) *ACORN Users' Guide, 1988*, London: CACI Ltd.
6 Metropolitan Borough of Bury, Aspinwall & Company Ltd and Department of Trade and Industry (DTI) Warren Spring Laboratory (1986), *A Research Study of the Effects of the Introduction of Continental-Style Wheeled Containers on the Generation and Collection of Household Waste*', Institute of Wastes Management.
7 Davies and Marsh (1986) op. cit.
8 Metropolitan Borough of Bury *et al.* (1986), op. cit.

Chapter 6

Civic amenity waste disposal sites: the Cinderella of the waste disposal system

P.C. Coggins, A.D. Cooper and B.W. Brown

INTRODUCTION

A characteristic of modern society is its propensity to generate more waste in quantity and in greater variety. This includes both industrial and household waste. Waste generated by industry and certain aspects of waste recycling are covered in geographical studies of resource management but geographers have had little involvement with household waste. Research in this area has been limited, and carried out mainly within the waste management sector, either at national or local level. Recent concern with the environment and waste recycling has brought the issue of household waste into the public domain, and proposals for new legislation are likely to have considerable impact on both management and policy issues.

The Public Health Act 1936 required that councils in the UK must remove house refuse from all houses without making a charge. The Control of Pollution Act 1974 reinforced this duty of councils to collect household waste, although those sections defining household wastes to be collected free of charge, and prescribed household wastes, were not implemented until 1988. Prescribed household waste, for which a charge may be made, includes garden waste and bulky items. Recent estimates suggest the annual total household waste generated in the United Kingdom probably exceeds 20 million tonnes, 10–11 kilos per household per week (CIPFA 1988a). This compares with 17 kilos in the 1930s, and includes substantially less ashes but more putrescible items, paper and plastics.

After local government reorganization in 1974, waste management in England was handled at two levels. District councils (or metropolitan boroughs) were waste collection authorities and shire counties (or metropolitan boroughs, singly or in local groups) waste disposal authorities. In Wales both functions were handled by the Unitary Authorities. The Environmental Protection Bill (Green Paper) of December 1989, enacted late in 1990, separated the waste regulation and waste disposal functions in England. Waste disposal authorities became waste regulation authorities and 'arms length' local authority waste disposal companies (LAWDACS) were established to handle waste disposal functions. Waste collection became the responsibility of local authority waste collection authorities.

Problems with the dumping of abandoned cars and bulky household items in the countryside led to the Civic Amenities Act in 1967 (Civic Trust 1967) and under Part III, civic amenity sites, or household waste sites were to be set up to receive such waste items from householders free of charge. Referred to in the Control of Pollution Act 1974, these sites became the responsibility of waste disposal authorities in England and the Unitary Authorities in Wales following local government reorganization in that year.

The 1978 Refuse Disposal (Amenity) Act consolidated the previous legislation related to such sites. The 1989 Green Paper, referred to above, proposed to transfer responsibility for these sites in England from the waste disposal authorities to the waste collection authorities, but following submission of comments on the Green Paper, it was announced in August 1989 that civic amenity sites are to remain with the present waste disposal authorities, who will seek tenders from the local authority waste disposal companies or the private sector for site provision.

Since 1974 the quantity of waste received at these sites has trebled to more than 4 million tonnes, and this now represents as much as 20–25 per cent of total household waste arisings. Whilst trends in household waste (per household) handled by waste collection authorities have remained stable, or even declined, over the last ten to fifteen years, civic amenity waste has experienced rapid growth, particularly in garden waste and bulky items. Where larger wheeled bins have been introduced for household waste, however, growth rates for collected household waste of 20–30 per cent per annum have not been uncommon.

These bins were first used in continental Europe, and arrived in the UK in 1980. The smallest size, 120 litre, is similar to the 90-litre ordinary dustbin, but the most common size is 240 litres, and some authorities offer 330 litre and even 500 litre sizes. At present over seventy waste collection authorities have introduced them. As will be discussed in the paper, the introduction of such wheeled bins raises a number of important issues both for waste collection and recycling.

Although civic amenity sites were initially intended to be waste disposal sites *per se*, many sites are now associated with waste recycling schemes (County Surveyors' Society 1987). As much as 10–15 per cent by weight may be recovered at such recycling sites, thus reducing the quantity for final disposal. With wheeled bins, however, householders may put some of their potentially recyclable items (paper, cardboard, metals, glass) into the wheeled bins. In addition, householders may make fewer visits to civic amenity sites, as other waste may also be put in the wheeled bin. Many local authorities have established recycling centres during the last ten years, ranging from solitary bottlebanks to clusters of containers for glass, paper, cans and textiles. Usually set up in connection with the recycling industry, and with proceeds going to charity, these represent another outlet for household recyclables which might be affected adversely by wheeled bins. Announcements in October 1989 indicate that the government intends that local authorities should aim to recycle up to 50 per cent of waste, with civic amenity sites and recycling centres presumably playing an important role.

RESEARCH AT LUTON COLLEGE OF HIGHER EDUCATION

Since 1983 research at Luton has focused on the public use of civic amenity waste disposal sites, including user habits, waste taken to such sites and the actual and potential role of waste recycling. Research funded by the Department of the Environment between 1984 and 1986 was concerned with six sites in England (Ampthill, Aylesbury, Derby, Hove, Northampton and Shoreham) and over 12,000 interviews were completed with site-users during an eighteen-month period including surveys on weekdays and weekends (Coggins *et al.* 1986). Between 1986 and 1988 further funding from the Department of the Environment focused on the potential impact of large-capacity household wheeled bins on waste inputs to civic amenity sites, including any impacts on waste recycling. For this, the research was based in Nottingham, with over 4,000 completed questionnaires (Coggins, Cooper and Brown 1988). In both these contracts the typical sample of site-users during a survey day exceeded 75 per cent. During these two projects Warren Spring Laboratory (WSL) Stevenage carried out waste category analyses at Aylesbury, Derby and in Nottingham (Barton *et al.* 1987; Poll 1988,

1989). In this work a smaller sample of users had their waste weighed by individual categories, and related to volume estimates. In Nottingham, waste from a relevant household refuse collection round was taken to Stevenage for detailed category/weight analysis on a regular basis to complement on-site surveys.

In addition to these surveys, smaller surveys have been conducted in Luton and in selected London boroughs. Based on all this research a summary paper on site design, location and management was presented early in 1988 (Coggins, Cooper and Galena 1988).

A further project, again funded by the Department of the Environment and extending to 1991, will look at the overall effectiveness of civic amenity sites and complementary waste services within the broader overall context of waste management. Preliminary results looking at the public use of civic amenity sites and recycling centres (Coggins, Cooper and Brown 1989) will be complemented by further fieldwork and household surveys, together with a detailed evaluation of costs and benefits entailed in operating such facilities. This will involve further co-operation with Warren Spring Laboratory and also with Midland Environment Ltd (MEL) of Aston Science Park, Birmingham, who have conducted similar research on civic amenity sites in the West Midlands (MEL Consumer Market Research 1987, 1988). In July 1989 the consortium was awarded the contract to monitor and evaluate the progress of the Sheffield Recycling City Project, set up by UK 2000 and Friends of the Earth.

SURVEY RESULTS

From the two research projects completed to date, a considerable quantity of information has been assembled about public use of civic amenity sites. Some 20,000 completed questionnaires, from fourteen different sites in England, covered daily and seasonal variations over five years.

Characteristics of the typical site-user will be summarized first in this section, and then the emphasis will be on the research in Nottingham (see Figure 6.1):

- The typical site-user is male, usually well over 85 per cent, with females more prominent in weekday surveys. (It is normal for about 50 per cent of site visits to be made at weekends.)
- Over 50 per cent of site-users are in the 25–44 age group, again with older people (over 65) tending to visit sites on quieter weekdays.

- Nearly 90 per cent arrive by car, site location as well as the quantity and type of waste being delivered requiring access to a vehicle. Vans and pick-ups are more common where more trade waste is deposited illegally, whilst urban sites attract users on foot.
- The majority of site-users bring the equivalent of a car boot full of waste (1–4 sack equivalents). Smaller quantities are brought by those arriving on foot whilst larger van and trailer loads (up to 500 kg) are not uncommon.
- Some 40 per cent of users quote weekly/fortnightly visits, and although this varies during the year, it indicates regular usage patterns.
- Reinforcing this last point, over 95 per cent of users say they have used the site before, and over 85 per cent say they do not use any other sites.
- Over 65 per cent of site-users make single purpose trips to the site. This may fall to less than 33 per cent on weekdays when people visit the site to or from work or shopping.

From traffic counter data most sites exhibit peak flows during late morning, and perhaps a secondary peak during late afternoon. Peak flows of 300–400 per hour are not uncommon, with random surges in vehicle numbers increasing queuing. This becomes particularly acute where users are unloading bulky items or large volumes of waste.

The site catchment area is related to road accessibility, car-ownership patterns (or access to alternative vehicles), and the existence of competing sites. Such catchment areas may be highly localized as with small urban or rural sites (for example, Shoreham, Ampthill), or very extensive where a large urban area may be served by only one site, as at Derby. In our research we have used postcodes to map catchment areas. Many local authorities talk of planning catchment areas with a five-mile radius, and this has been confirmed in most of our survey work, with often 80 per cent of users coming from within two and three miles of the site.

The contrasting catchment areas for two of our Nottingham sites are shown in Figure 6.2.

Interviewers noted the types of waste brought to the sites, and since 1985 they have also estimated these in terms of sack equivalent. Warren Spring Laboratory analysed and weighed site-users' waste on a sample basis (18 per cent of site-user surveys in Aylesbury and

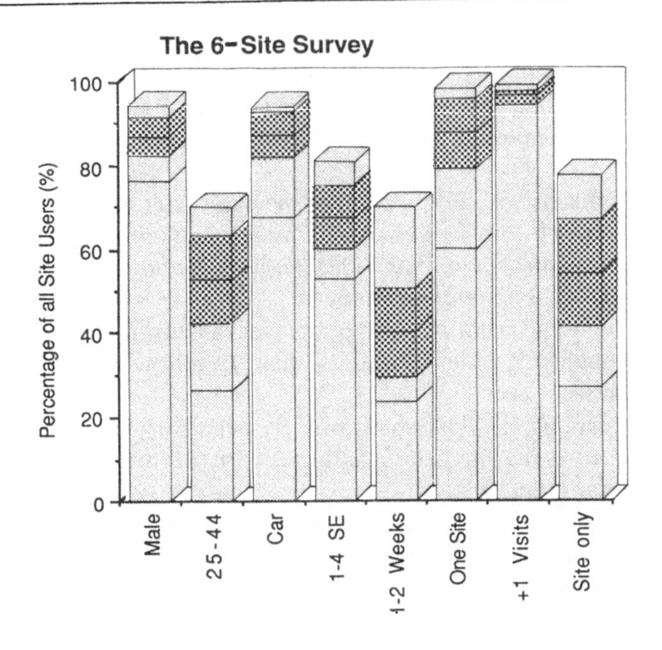

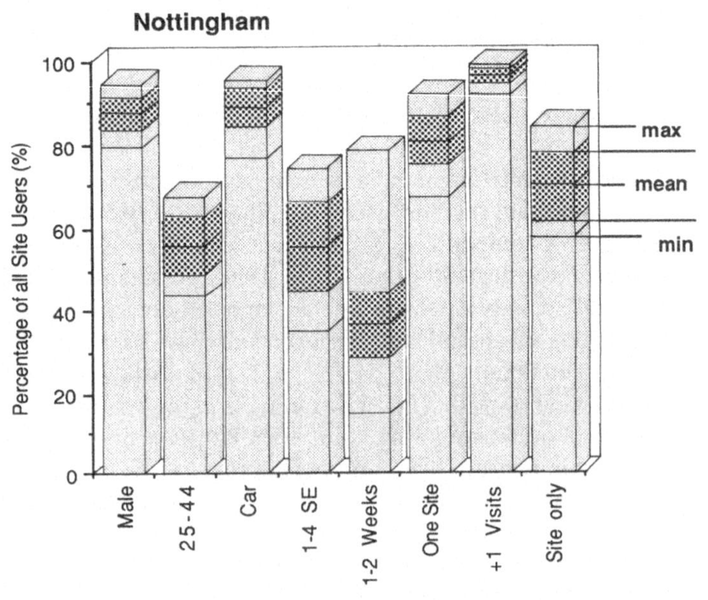

Figure 6.1 The typical civic amenity site user

Derby, and 12 per cent in the Nottingham research). The waste classification was modified during the first research project, and only the last three survey batches in 1985 are directly comparable with the Nottingham data. Garden waste (both small and large) is normally brought by well over a third of site-users, rising to over 80 per cent in spring and autumn. A third bring ordinary household waste, a third bring paper and cardboard, and nearly a third bring 'do it yourself' (DIY) (both combustible and non-combustible) and sawn timber. Bulky household (household large) items are brought by 10–15 per cent of users.

For recyclable products 5–7 per cent brought bottles and glass, 10 per cent brought metals, only 1–2 per cent brought motor oil. Such recyclable products are brought by more people where recycling is actively promoted. At Camden in London, for example, over 40 per cent of users brought bottles and glass, 70 per cent brought paper, 20 per cent brought cardboard and 25 per cent brought metals. A surprising proportion bring ordinary household waste (household small), as they generate too much for the weekly bin collection.

Figure 6.3 shows monthly averages for the garden waste component. Spring and autumn peaks are apparent from these data, averaged over a three-year period, thus depressing the annual changes apparent in Figure 6.4 which brings together the two sets of mean summary data of waste deposited during the November and January surveys.

WHEELED BINS: IMPLICATIONS FOR SITE USE

The focus in the Nottingham research was to identify any changes resulting from the introduction of wheeled bins (Jones 1988). Three civic amenity sites were studied, one in south Nottingham and two in adjacent districts, serving a total population of over 300,000. Each site received over 5,000 tonnes of waste in 1987–8 (excluding recovered materials, estimated by the waste disposal authority to be an additional 10 per cent by weight), and this represented between 9 per cent and 15 per cent of total household waste in the three districts. Two additional civic amenity sites within one of the districts were not covered, but users of a temporary Saturday skip facility were surveyed until the implementation of wheeled bins led to their withdrawal. Warren Spring Laboratory carried out sample weighing on-site, and also monitored household waste before and after the introduction of wheeled bins.

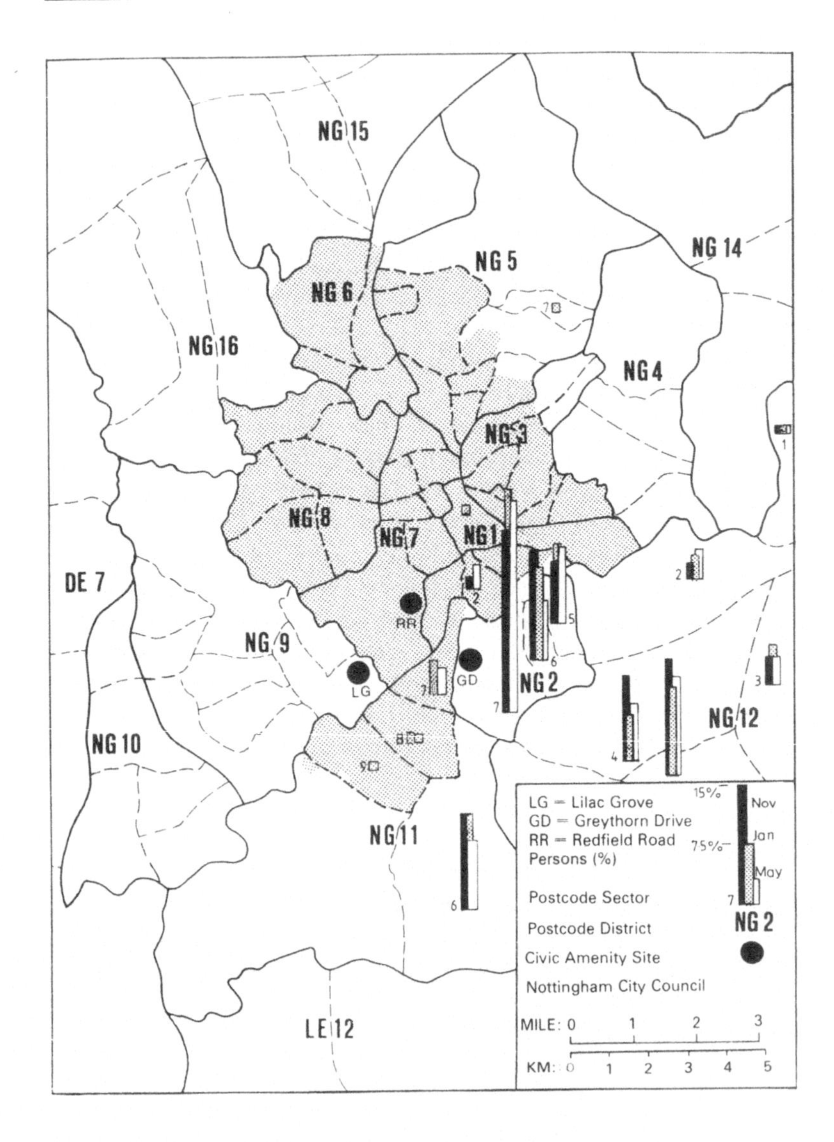

Figure 6.2a Site users' home postcode: Graythorn Drive, Nottingham, 1987–8

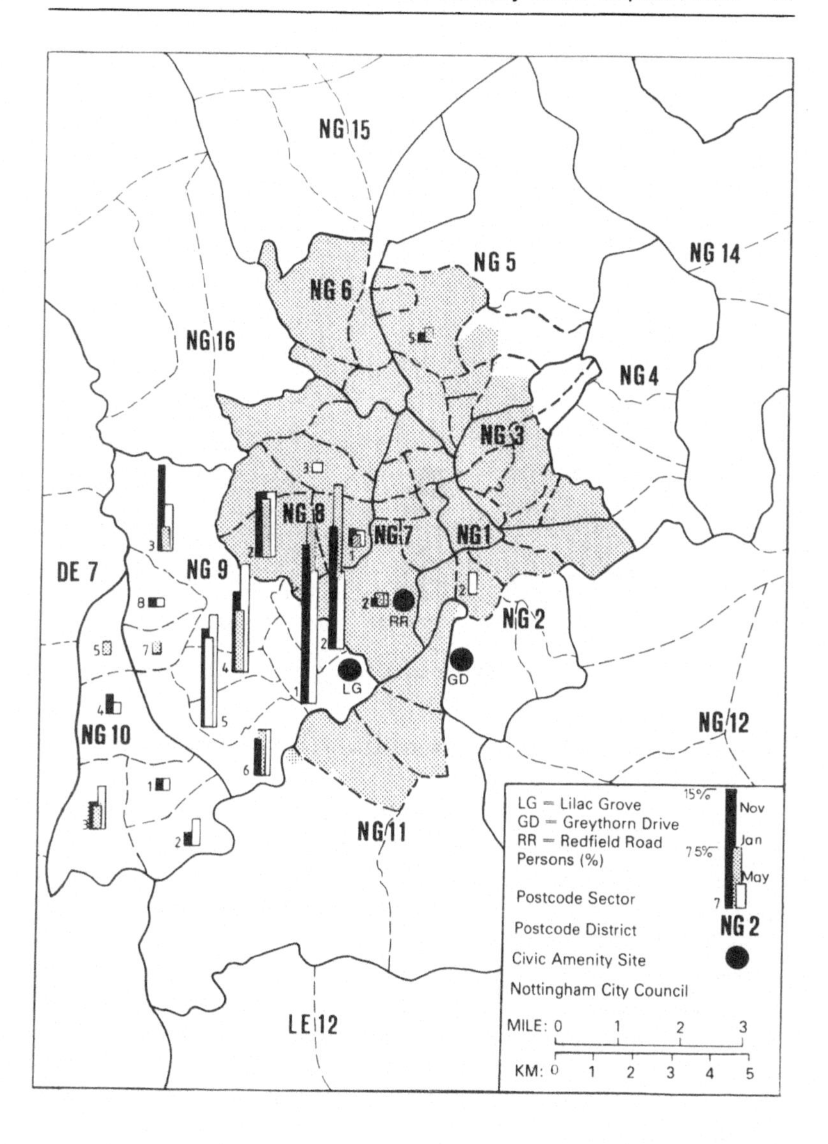

Figure 6.2b Site users' home postcode: Lilac Grove, Nottingham, 1987–8

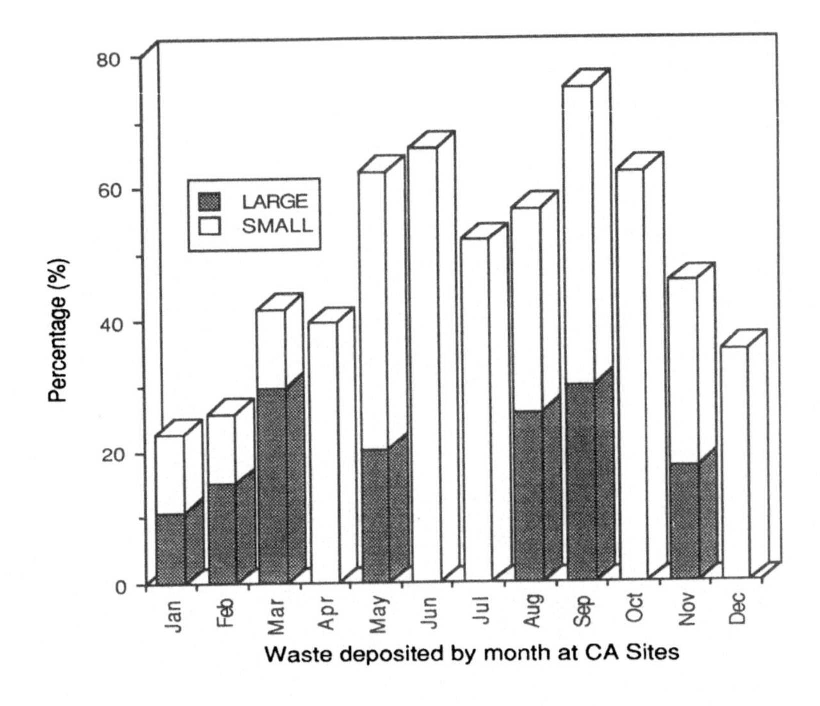

Figure 6.3 Garden waste by season (based on mean values, 1985–8)

By weight, site-users brought, on average, 50 kilos compared to 44 kilos at Aylesbury and Derby. Garden waste accounted for over 40 per cent of this, rising to over 50 per cent in spring and autumn. DIY accounted for 17 per cent, increasing to 25 per cent if all wood is included in this category.

Ordinary household waste accounted for nearly 10 per cent, and bulky household waste 6 per cent. Of recyclable waste, paper and cardboard represented 5 per cent and metals nearly 10 per cent. As might be expected, over 40 per cent of weight of the waste was larger than 450 mm, and thus could not be put in the conventional dustbin.

Following the introduction of wheeled bins in south Nottingham, people supplied with one who continued to use civic amenity site tended:

- To bring larger loads by weight (52 kilos, compared to 47 kilos) and by volume (including more oversize items).

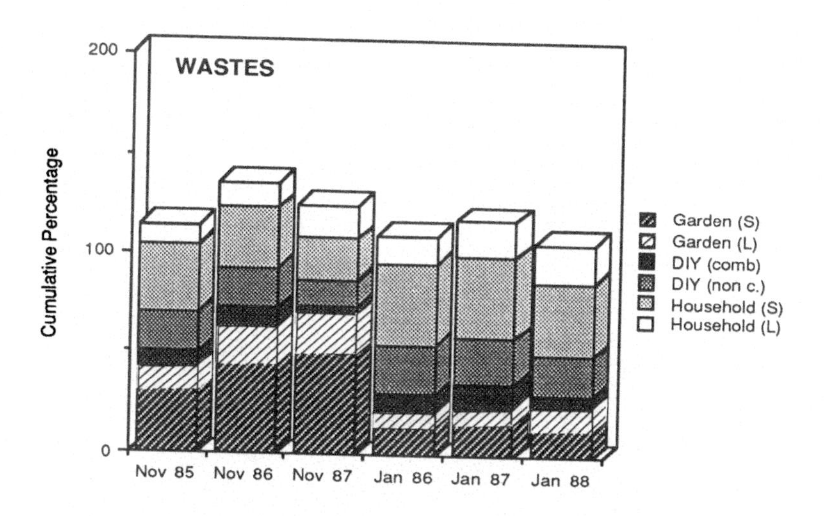

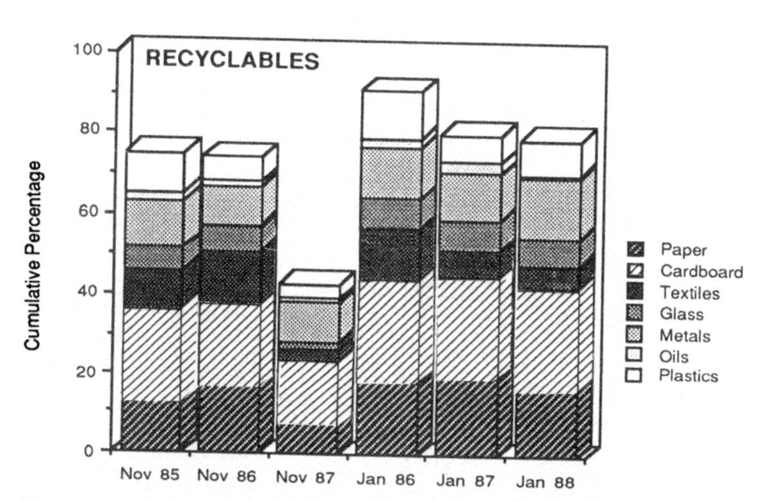

Figure 6.4 Waste and recyclable products, by percentage, deposited at civic amenity sites

- To use civic amenity sites a little less frequently, monthly being more common than weekly/fortnightly.
- To bring less small garden waste, paper and cardboard. In contrast, they bring more DIY (including timber), household large items and metals.
- To bring several different categories of waste.

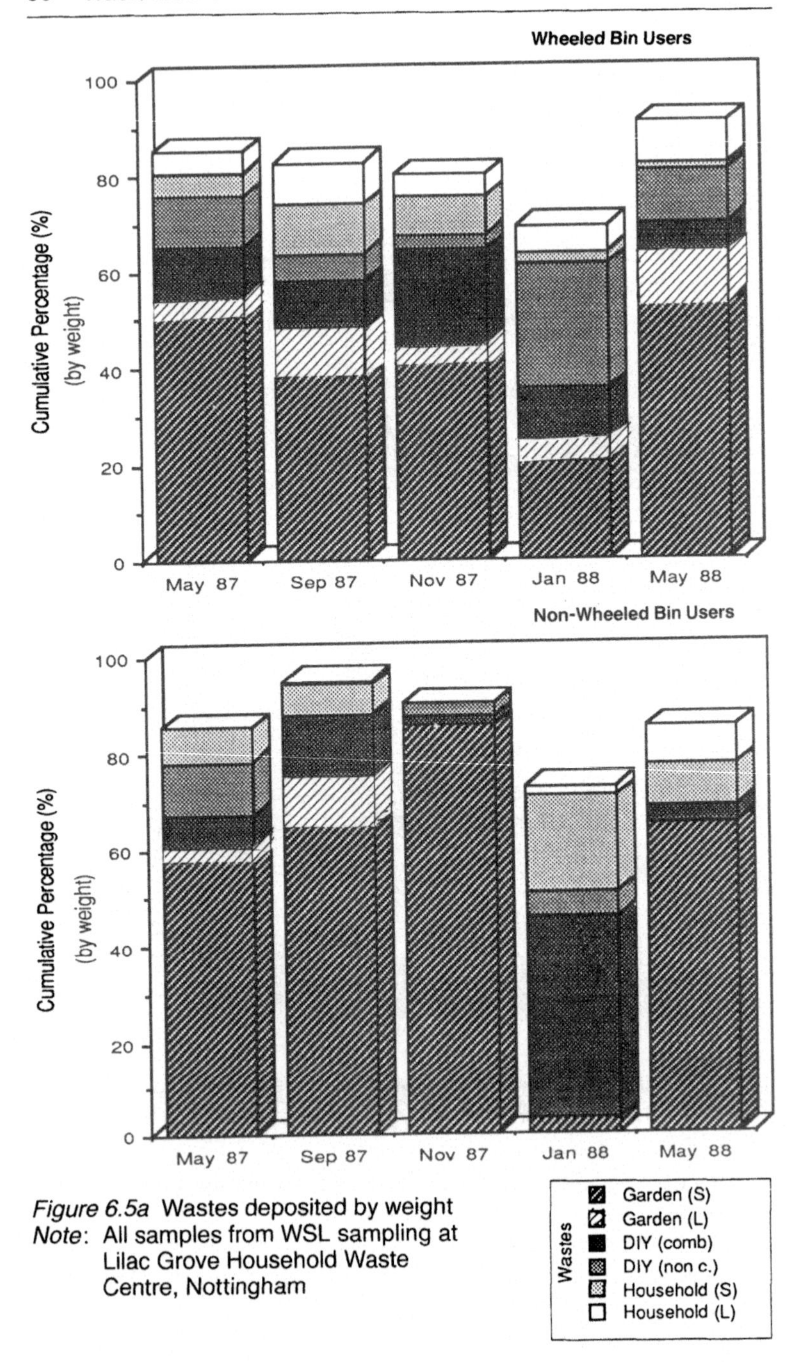

Figure 6.5a Wastes deposited by weight
Note: All samples from WSL sampling at Lilac Grove Household Waste Centre, Nottingham

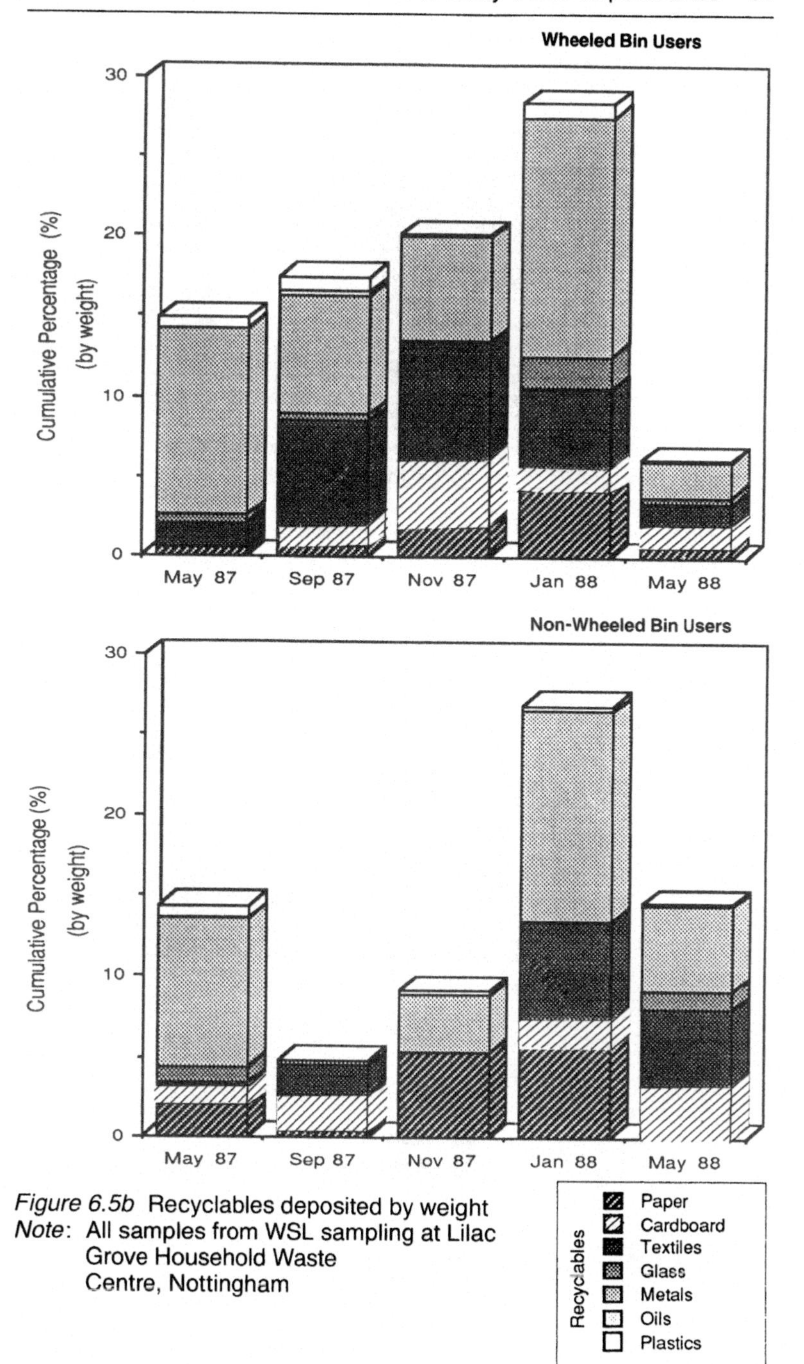

Figure 6.5b Recyclables deposited by weight
Note: All samples from WSL sampling at Lilac Grove Household Waste Centre, Nottingham

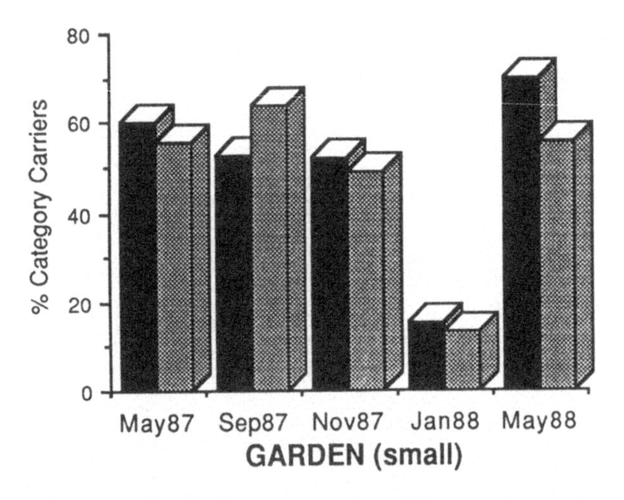

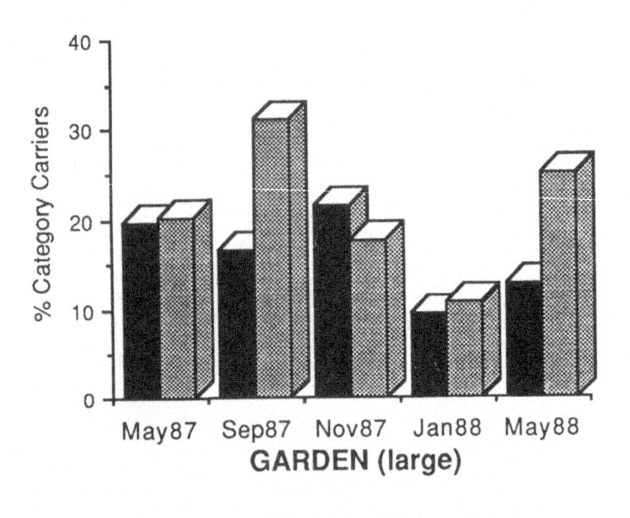

Non-WB Users
WB Users

Figure 6.6 Civic amenity wastes/recyclables deposited by the percentage of category carriers

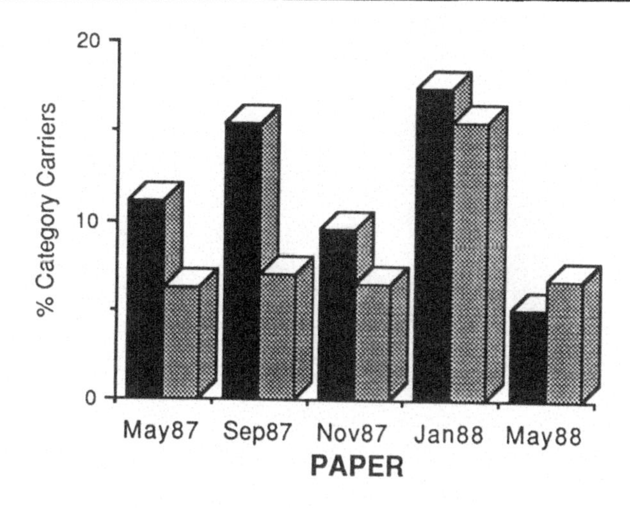

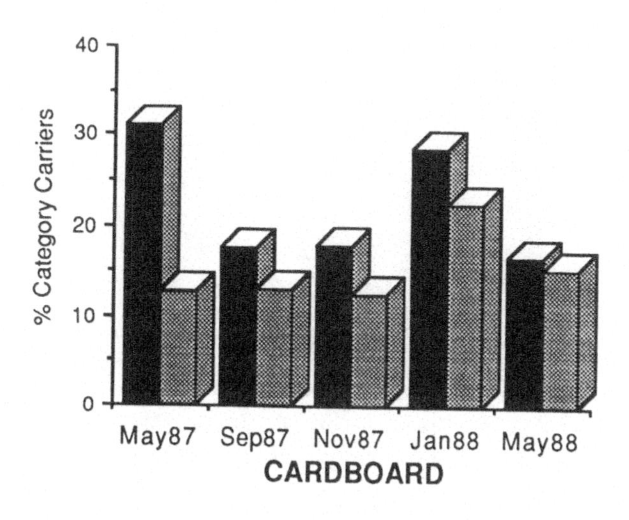

Non-WB Users
WB Users

Figure 6.6 (continued)

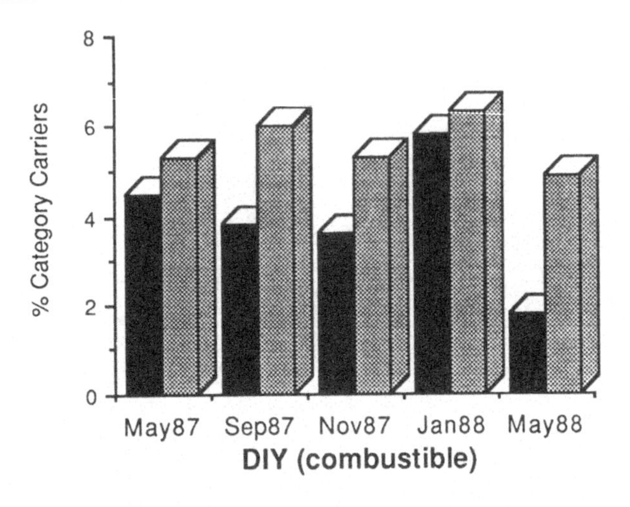

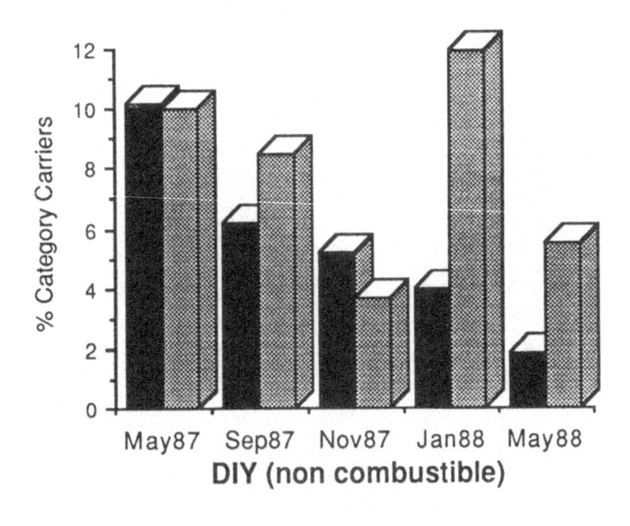

Figure 6.6 (continued)

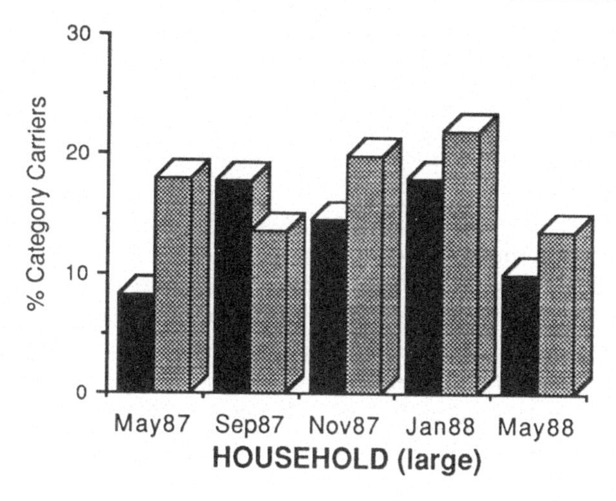

Non WB Users

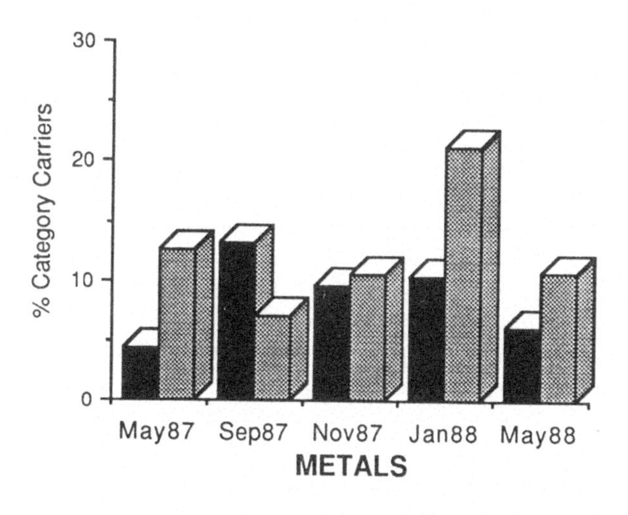

■ **Non-WB Users**
▩ **WB Users**

Figure 6.6 (continued)

This implies both a diversion of small waste into the wheeled bin and continued use of civic amenity sites for certain categories of waste.

Figure 6.5 compares wheeled bin users and non-wheeled bin users in the WSL sample (proportion by weight), whilst Figure 6.6 uses the total questionnaire sample to illustrate the above points, based on users bringing selected waste categories. Overall trends are apparent, although the November 1987 and January 1988 data in Figure 6.5 for non-wheeled bin users are based on a very small sample. In contrast, following the introduction of wheeled bins, the quantity of weekly household waste increased by over 50 per cent from 11 kilos to 18 kilos, with more garden waste (especially small) paper and cardboard and small DIY material. Figure 6.7 summarizes these data, as analysed by WSL and the proportions are based on weight.

IMPLICATIONS OF RECENT CHANGES IN WASTE MANAGEMENT

It is apparent that introducing wheeled bins will have impacts on household waste collection, and consequently on civic amenity waste inputs (Mansfield 1987; Metropolitan Borough of Bury 1986). The increase of over 50 per cent in household waste collected via the wheeled bins in Nottingham will pose problems if this is sustained, especially if there are strong seasonal variations related to the gardening calendar. This waste then has to be disposed of by Nottinghamshire County Council, with potential pressures on landfill void space and disposal costs. Net collection costs per household in Nottingham in 1986–7 were £27 per tonne (CIPFA 1988a), and the reduction in labour costs using wheeled bins will be offset by this increased tonnage. An aim of the present research contract is to establish the costs and benefits of civic amenity site operations, a task complicated by the considerable variety of accounting procedures used by local authorities. The sensitivity of this type of data has been highlighted in 1989 by the decision by CIPFA not to publish its annual survey of waste collection and waste disposal, given the potential problems caused by the requirements of competitive tendering.

At civic amenity sites it is the reduction in potentially recyclable items that is the most important trend. Estimates in Nottinghamshire quote some 10 per cent by weight extracted by the site

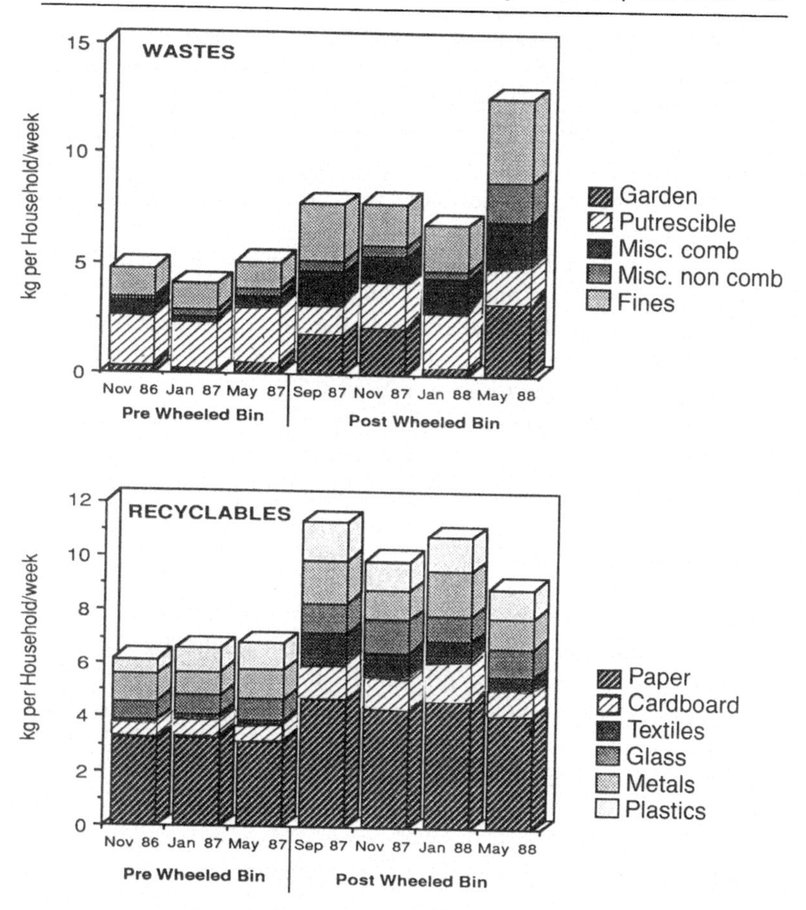

Figure 6.7 Domestic waste analysis

operators working as licensed totters for salvage/recycling, and this could decline as paper, cardboard, glass and small metal items are put in wheeled bins. However, such bulk recycling may only be one component in the licensed totters' income as in Nottinghamshire they also have a well-established network of auction rooms to sell furniture, electrical goods, bric-à-brac and other saleable household waste items.

In both cases the issue of privatization is relevant. At the waste collection level, competitive tendering may lead to an expansion of wheeled bin schemes and savings in manpower. On the other hand increasing quantities of waste may arise, unless charging for garden waste (under the 1988 regulations) is introduced. At civic amenity sites, leasing them to licensed totters may be less attractive if the potential waste inputs for recycling are reduced. In waste management terms, these entail potentially new stresses in providing the notionally free public services, with maximum efficiency and minimum cost.

In environmental terms such changes may lead to greater inputs into landfill, and less materials being recycled. As landfill sites are filled and returned to other uses, the trend towards more independent civic amenity site locations will increase and proposals for site relocation may create a NIMBY (Not In My Back Yard) reaction from the general public. Except in the most sensitive locations, civic amenity sites are unlikely to be candidates for environmental assessment (Department of the Environment 1988).

CONCLUSIONS

A considerable quantity of data has been assembled on the types of waste brought to civic amenity sites, its weight by individual category, and volume by individual category. One aspect of the research has been to link these measures together, and to use them to predict gross waste inputs by individual category. Table 6.1 summarizes the weight per sack equivalent data for the WSL sample surveys, with the distinction between wheeled bin and non-wheeled bin users evident. Given the wide range for some categories it was necessary to modify the data for regression purposes. As an example, various regressions for garden (small) waste are shown in Figure 6.8. Garden (small) is usually the single most important item brought to civic amenity sites, and this waste category is used to illustrate both the problems of predicting potential waste inputs, and how removing extreme values (one user brought 500 kg) may allow the use of simple regression without sacrificing reliability. Summary values for all waste types are given in Table 6.2.

In Nottingham, these data have been used in order to try and predict weekly tonnages of waste inputs into civic amenity sites. Table 6.3 compares various estimates of total weekly inputs with actual weekly tonnages. Given the problems inherent in using questionnaire sample data, estimating sack equivalent, WSL sample weight data, traffic counter inaccuracies and the reliability of actual tonnage data, the 'best fit' estimates are remarkably consistent for

Table 6.1 Weight (kg) per sack equivalent values from WSL surveys (Nottingham: WSL sampled surveys)

Category	*n*	*x*	*STD*	*n*	*x*	*STD*	*n*	*x*	*STD*
Garden (S)	97	11.1	7.3	102	11.5	7.5	199	11.3	7.4
Garden (L)	30	5.1	3.0	15	5.5	3.2	45	5.2	3.1
Paper	13	10.2	6.8	27	8.8	4.9	40	9.3	5.6
Cardboard	40	2.1	1.0	47	2.8	1.8	87	2.5	1.5
Timber	36	8.8	4.9	28	11.4	6.3	64	9.9	5.7
Textiles	15	6.0	2.6	19	6.1	2.8	34	6.1	2.7
Bottles/glass	10	11.7	6.5	17	13.7	6.9	27	12.9	6.8
Plastic film	6	3.0	1.5	4	1.3	0.7	10	2.3	1.5
Dense plastic	10	2.6	2.4	18	2.9	2.6	28	2.8	2.5
DIY (comb.)	16	8.7	4.5	24	7.4	5.3	40	7.9	5.0
DIY (non-comb.)	25	27.5	22.2	29	28.8	28.4	54	28.2	25.7
Motor oil	3	14.8	11.4	3	10.0	3.5	6	12.4	8.4
Household (S)	43	6.6	4.1	86	7.6	4.8	129	7.2	4.6
Household (L)	40	7.6	4.3	30	7.8	7.7	70	7.7	6.0
Metals	51	8.3	6.9	33	10.5	11.8	84	9.1	9.2

n Sample size
x Mean
STD Standard deviation

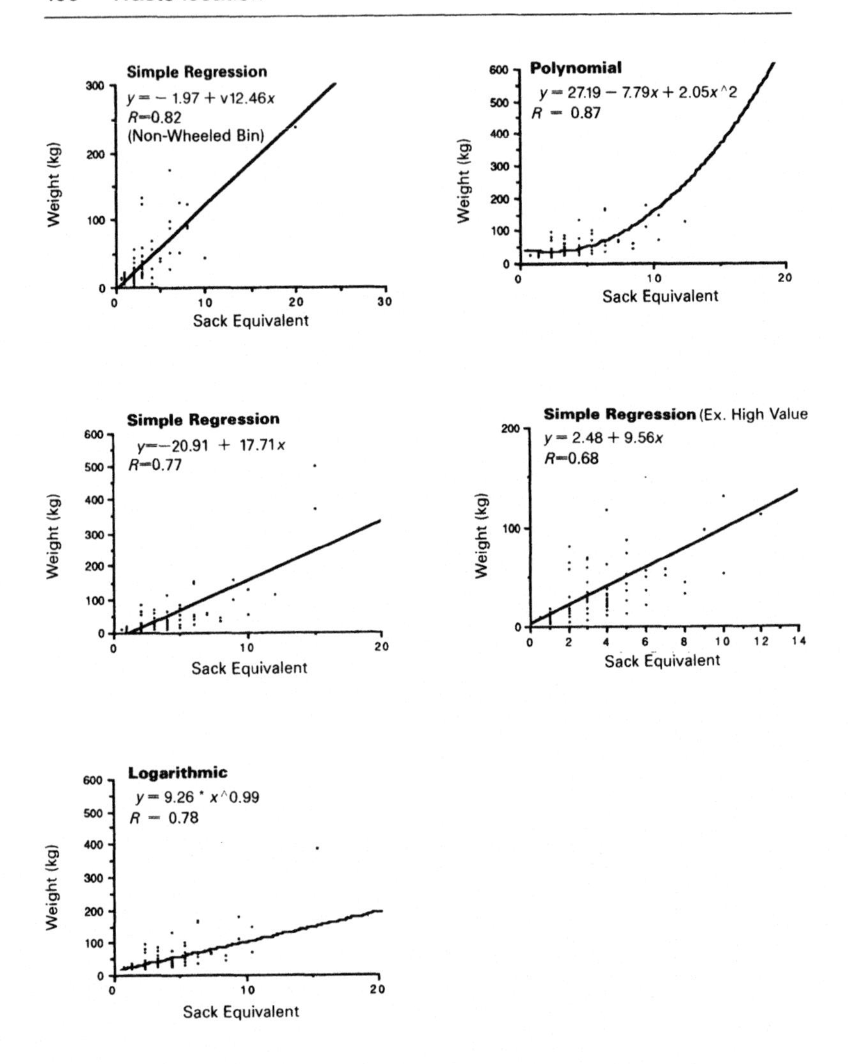

Figure 6.8 Regression analysis – garden waste (small) deposited at Lilac Grove CA site, Nottingham, by weight

Note: All wheeled bin users unless otherwise stated

the Greythorn Drive site. This methodology is to be further investigated in the new research project. Analysis of various alternatives suggests that seasonal site estimates, and average weekly tonnages over a one- to three-month period tend to give the closest fit.

Table 6.2 Weight per sack equivalent values (kg) from regression analysis

Category		*WSL*	*r*	*ex*	*x*
Garden (S)	NU	11.5	10.5	9.9	10.6
	U	11.1	+3.2	12.0	11.6
Garden (L)	NU	5.5	6.4		6.0
	U	5.1	−0.3	4.2	5.1
Paper	NU	8.8	8.3		8.6
	U	10.2	11.1	9.8	10.7
Cardboard	NU	2.8	2.8		2.8
	U	2.1	2.4		2.3
Timber	NU	11.4	10.3	9.5	10.9
	U	8.8	7.7	7.3	8.3
Textiles	NU	6.1	5.8	5.9	6.0
	U	6.0	6.0	3.5	6.0
Glass	NU	13.7	11.3	12.0	12.5
	U	11.7	15.0		13.4
Plastic film	NU	1.3			1.3
	U	3.0			3.0
Dense plastic	NU	2.9	4.9		3.9
	U	2.6	2.0		2.3
DIY (comb.)	NU	7.4	5.9		6.7
	U	8.7	8.6		8.7
DIY (non-comb.)	NU	28.8	25.9	24.6	26.4
	U	27.5	25.3	14.5	26.4
Motor oil	NU	10.0			10.0
	U	14.5			14.5
Household (S)	NU	7.6	7.7		7.7
	U	6.6	5.7		6.2
Household (L)	NU	7.8	11.3	11.5	9.6
	U	7.6	6.0	6.0	6.5
Metals	NU	10.5	11.8		11.2
	U	8.3	7.6	7.3	8.0
Other	NU	9.1			9.1
	U	8.9			8.9

Notes:
NU Non-wheeled bin user.
U Wheeled-bin user.
WSL Mean weight per sack equivalent (SE) calculated from WSL data.
r Simple regression calculated from WSL weight/SE data.
ex Modified simple regression of WSL weight/SE data by excluding unusually extreme high values.
x Mean weight/SE values calculated from *WSL*, *r*, and *ex*, values but EXCLUDING extreme values.

Table 6.3 Estimated weekly tonnages from civic amenity sites

Date	*Site*	*Traffic*	*'Best'*	*Range*	*Week*	*3 wks*	*Qtr*	*FIT*
11/86	GD	1,893	99	78–99	115	102	*99*	100.0
	LG	1,659	68	56–68	101	*93*	95	73.1
	RR	1,333	63	50–64	*63*	69	70	100.0
01/87	GD	1,065	46	36–48	16	*45*	65	102.2
	LG	1,034	31	26–33	*32*	39	67	96.9
	RR	4,068	110	110–137	35	47	*66*	166.7
09/87	GD	2,915	114	108–128	213	*114*	93	100.0
	LG	2,119	80	75–80	229	177	*136*	58.8
	RR	1,720	76	73–76	167	127	*114*	66.7
11/87	GD	2,074	63	63–91	50	*63*	64	100.0
	LG	1,405	49	41–49	120	139	*100*	49.0
01/88	GD	2,086	59	43–59	88	89	*81*	72.8
	LG	1,129	34	34–41	*24*	139	56	141.7
05/88	GD	3,541	140	107–145	109	*141*	134	99.3
	LG	2,329	93	77–93	134	145	*123*	75.6

GD Greythorn Drive CA Site, Nottingham
LG Lilac Grove CA Site, Nottingham
RR Redfield Road CA Site, Nottingham
Traffic Estimated no. of vehicles using site
'Best' Closest ESTIMATE to ACTUAL
Range Estimates based on Site SE, Monthly SE (SE for all sites/same date) and SE (all sites/all dates)

Week ACTUAL tonnage for corresponding week
3 wks ACTUAL tonnage based on 3-weeks average
Qtr ACTUAL tonnage based on 13-weeks average
141 ACTUAL tonnage closest to estimate
FIT % closeness of fit (Estimate/Actual*100)

A second research topic to be investigated over the next two years is to look at site catchment areas in more detail, including household-based survey work, and further work at selected civic amenity sites. This will involve examining waste generation activities of site-users and non-users, and reasons why people use, or do not use, civic amenity sites, together with site accessibility and management policies.

Third, the overall cost-benefit analysis of civic amenity sites within the waste management system will be examined. Whether they have reduced fly-tipping by householders will be included in the research. The public who use sites see them as a valuable public service, they are regular users, and are happy to bring recyclable items and to assist the operative in segregating these items from waste. In so doing they subsidize the costs of assembling recyclable items, and the larger quantities available at civic amenity sites provide more scope for negotiating more stable contracts with recycling companies.

NOTE

The views in this chapter do not necessarily reflect the views of the Department of the Environment, nor the local authorities mentioned.

REFERENCES

Barton, J.R., Wheeler, P.A. and Poll, A.J. (1987) *The Effects of the Introduction of Large Wheeled Bins on the Composition of Domestic Waste*, paper to Institute of Wastes Management, Buxton.

Carter, J.B. (1984) *Dustbins on Wheels – the Rolling Programme Starts*, paper to Institute of Wastes Management, North-West Centre.

Chartered Institute of Public Finance and Accountancy (CIPFA) (1988a) *Waste Collection Statistics 1986–1987 Actuals and Estimates*, Statistical Information Service.

Chartered Institute of Public Finance and Accountancy (CIPFA) (1988b) *Waste Disposal Statistics 1986–1987 Actuals and Estimates*, Statistical Information Service.

Civic Trust (1967) *Disposal of Unwanted Vehicles and Bulky Refuse*, London.

County Surveyors' Society (1987) *Household Waste, the Business of Reclamation*

Coggins, P.C., Cooper, A.D. and Brown, R.W. (1988) *Nottingham Surveys: The Impact of Wheeled Bins* (for Department of the Environment), Luton.

Coggins, P.C., Cooper, A.D. and Brown, R.W. (1989) *Civic Amenity Sites and Recycling Centres: Complementary or Competing Facilities*, paper to Institute of Wastes Management, London Centre, autumn meeting.

Coggins, P.C., Cooper, A.D. and Cole, M. (1986) *Characteristics for Recycling at Civic Amenity Sites. Final Report and Conclusions: Ampthill, Aylesbury, Derby,*

Hove, Northampton and Shoreham (for Department of the Environment), Luton.

Coggins, P.C., Cooper, A.D. and Galena, M. (1987) 'When your rubbish gets a set of wheels', *Surveyor*, 18 June, 17–18.

Coggins, P.C., Cooper, A.D. and Galena, M. (1988) 'Civic amenity sites: parameters for site location and management', *Wastes Management* 78 (5), 297–314.

Department of the Environment (1988) *The Town and Country Planning (Assessment of Environmental Effects) Regulations 1988*, London: HMSO.

Department of the Environment (1989a) *The Role and Functions of Waste Disposal Authorities: A Consultation Paper*, London: HMSO.

Department of the Environment (1989b) *The Role and Functions of Waste Disposal Authorities: Announcement of Government Decisions*, London: HMSO.

Jones, D. (1988) 'Cleansing services – meeting today's challenges', *Wastes Management* 78 (11), 748–56.

Mansfield, R. (1987) *Wheeled bins – the Bolton Way*, paper to Institute of Wastes Management, North-East Centre.

Metropolitan Borough of Bury, Aspinwall & Company Ltd and Department of Trade and Industry (DTI) Warren Spring Laboratory (1986) *A Research Study of the Effects of the Introduction of Continental-Style Wheeled Containers on the Generation and Collection of Household Waste*, Institute of Wastes Management.

Midland Environment Ltd (MEL) Consumer Market Research (1987) Questionnaire survey of people using public waste disposal sites, Birmingham.

Midland Environment Ltd (MEL) Consumer Market Research (1988) *Untapped Potential for Recycling at Civic Amenity Sites*, Birmingham.

Poll, A.J. (1988) *The Effect of the Introduction of Wheeled Bins on the Quantity and Composition of Domestic and Civic Amenity Waste in Nottingham*, Warren Spring Laboratory, CR 3087 (MR).

Poll, A.J. (1989) *The Effect of Wheeled Bins on Domestic and Civic Amenity Waste*, Warren Spring Laboratory, LR710 (MR) M.

Chapter 7

A geographic information systems approach to locating nuclear waste disposal sites

Steve Carver and Stan Openshaw

INTRODUCTION

This chapter is a revised version of a report submitted to UK NIREX Ltd as part of their 1988 public consultation exercise on radioactive waste disposal. It discusses the problems of radwaste disposal in the UK and makes suggestions as to how NIREX's interpretation of government radwaste policy and siting guidelines might be improved. Particular reference is made to the potential of Geographic Information Systems (GIS) and multi-criteria evaluation (MCE) methods used with existing digital map data bases as a means of both making and validating important siting decisions. The use of the visualization properties of GIS as a tool for explaining the siting process to non-expert audiences is also examined.

Great Britain, like other nuclear nations, has in the course of its civil and military nuclear power programmes, created a sizeable radioactive waste problem. This problem has only recently begun to receive wide-scale interest from the public, politicians and academics alike. In the last decade and a half a number of government reports and inquiries have drawn attention to radioactive waste. The sixth report of the Royal Commission on Environmental Pollution on 'Nuclear Power and the Environment' (1976) was the first formal recognition of the radwaste problem. Since then the Windscale Inquiry (Parker 1978), the Sizewell B Public Inquiry (Layfield 1987) and the first report of the House of Commons Select Committee on the Environment (1986) have all expressed concern over the increasing stocks of radwaste awaiting disposal and the scale of future production.

Radioactive wastes have been created as a result of nuclear power generation, research, industrial and medical uses of radioactive materials, and, much more significantly, as a result of the reprocessing of spent nuclear fuel. Table 7.1 gives a breakdown of waste arisings by source. It is now in the national interest that this waste should be stored or disposed of in a safe and acceptable manner. It is too late to argue that these wastes should never have been created in the first place. Given the size of waste stocks and projected future arisings, the limited capacity of current disposal and storage facilities means that a new radwaste facility needs to be commissioned by the early part of the next century, with the possibility of others in years to come. Table 7.2 shows estimated current stocks and future arisings of radioactive waste. At present low-level radioactive wastes (LLW) are disposed of at Drigg, near Sellafield in Cumbria. All intermediate-level (ILW) and high-level wastes (HLW) are stored on site or at Sellafield pending disposal or alternative storage arrangements.

Although, technically, radioactive waste disposal is not as complex as reactor and reprocessing technology, it is still an unproven science because of the very long time-scales involved. Some of the major radionuclides present in radwaste have extremely long half-lives and so need to be isolated from the environment for hundreds of thousands of years. This raises questions about both the necessity of nuclear power and, more specifically, whether indefinite storage is more appropriate than disposal for existing and future

Table 7.1 Sources of radioactive waste holdings in 1987 (%)

Source	*Waste category*		
	LLW	*ILW*	*HLW*
Sellafield	23	75	86
Other nuclear industry	40	14	0
Research, medical and industrial	36	11	14

Source: Radioactive Waste Management Advisory Committee (1988: 53).
Notes:
1 This table artificially separates 'other nuclear industry' from nuclear research which is included under 'research, medical and industrial' sources, thereby making it appear that a significant proportion of ILW and HLW is created outside of the nuclear industry. In fact, virtually all ILW and HLW is created by the nuclear industry and its research organizations.
2 The proportion of LLW arising from operations at Sellafield is massively underrepresented. This is due to most of Sellafield's LLW being disposed of at the nearby Drigg site.

Table 7.2 Estimated holdings and future arisings of radioactive waste

Waste type	*Holdings (1987)*	*Total arisings (2030)*
LLW	2,340	1,411,000
ILW	59,400	259,200
HLW	517	3,030

Source: Radioactive Waste Management Advisory Committee (1988: 51,53,55).
Notes: All figures are volumes expressed in cubic metres, 1987 figures are conditioned volumes and 2030 figures include decommissioning wastes.

wastes. More relevant, from a geographical point of view, is the question as to where such a facility should be located. These questions, coupled with heightened media attention, current environmental awareness, public fears and political manoeuvering, make radwaste an issue which is likely to dominate the nuclear debate over the next decade.

The scope of this chapter is not to question whether Britain should be producing the waste in the first place because, regardless of the UK's future nuclear policy, a major waste disposal problem already exists (see Table 7.2). Whatever policies are adopted in the longer term, there is already a substantial commitment to a major reprocessing programme which cannot now be easily halted. Moreover, should nuclear power ever be abandoned, then short-term waste arisings from the decommissioning of redundant nuclear plant could be overwhelming. The really important questions concern, therefore, not so much from where the wastes initially came and why they were produced, but rather what to do with them now that they exist and to what extent the future inventory can be minimized by sound management practices. Assuming the wastes are either to be stored or disposed of somewhere, then the key engineering question concerns the nature of the storage or disposal technology being used, whilst the key geographical question concerns *where* it should be stored or disposed. It is the latter question which is arguably the greatest source of public, and political, concern, and this is the question which this chapter addresses.

PAST AND PRESENT SITING PROCEDURES

The Nuclear Industry Radioactive Waste Executive (NIREX) was created in 1982 to implement government strategy for the disposal of radioactive waste. In particular it was charged with the responsibility for developing a new facility for the disposal of low and intermediate level radioactive wastes. In 1985 NIREX became UK NIREX Ltd, making it a separate legal entity and able to operate as a commercial organization. However, 99 per cent of the company shares are still held by its founding bodies – British Nuclear Fuels (BNFL), the Central Electricity Generating Board (CEGB), the South of Scotland Electricity Board (SSEB) and the United Kingdom Atomic Energy Authority (UKAEA). A token 1 per cent share is held by the Secretary of State for the Environment on behalf of the government. As a result NIREX technically remains the waste disposal arm of the nuclear industry.

The fact that NIREX is a private company means that it has no statutory responsibility under any Act of Parliament. This is significant because it places considerable stress on their need to demonstrate an overwhelming 'national interest' argument; that is, the waste has been created, and it is now in the national interest to find a safe place to bury it. This is important because NIREX is not a government department with a statutory duty of service to the country, and it means that NIREX must present a much stronger case in favour of any specific site than might otherwise be required. Therefore NIREX needs to be in a position to make the strongest possible case if it is to receive public and parliamentary approval. The task of NIREX is to find a site or sites for the disposal of low and intermediate level radioactive wastes. This has been made even more difficult from the outset by the lack of a comprehensive and coherent set of guidelines by which to implement the government's waste management policy and some doubts about precisely what the policy amounts to.

Draft disposal guidelines published by the Department of the Environment (DE) in 1985 stated that, in the evaluation of alternative sites, the developer (NIREX) '... will be expected to show that he has followed a rational procedure for site identification ... but he will not be expected to show that his proposals represent the best choice from all conceivably possible sites' (DoE 1985: 19). Earlier versions of the same document appeared more rigid, stating that the developer must demonstrate that in selecting a preferred site, better site options had not been ignored (DoE) 1983). This seemingly deliberate de-emphasis of the importance of site comparisons allows 'industry preferred' sites to be forwarded without a comprehensive and rational evaluation of the alternatives, and with no proof that even a relatively 'good' site has been found. There is a strong case for suggesting that, as a matter of good siting practice, any major developments which rely on national interest arguments really do require a level of comparative site evaluation which is far more rigorous than the current DoE recommendations would imply. When massively unpopular developments are being proposed, the responsibility should presumably lie with the developer, and not the protestor, to demonstrate in a scientific manner that the 'best' site and not merely a 'feasible' site has been found.

It is fairly obvious that the location of a repository depends on many factors in addition to safety and feasibility. However, convenience to NIREX cannot rank high on this list; neither should operational considerations such as transport costs and site ownership. The problem here is that DoE siting guidelines are little more than statements of general principle, that contain no details of specific siting criteria. The relevant factors of hydrogeology, population distribution and accessibility are merely listed, but their interpretation is left to NIREX. How these factors are to be used is undefined with no guidance as to how they can be applied in real siting situations.

Previous attempts to find a site to accept LLW and ILW have failed in the face of strong public and political opposition. Originally two sites were selected, one to take LLW and short-lived ILW in a shallow burial facility, the other to take longer-lived ILW in a deep underground repository. These were the CEGB's Elstow storage depot in Bedfordshire and a disused anhydrite mine at Billingham, Cleveland owned by ICI. Publicity on the site-selection procedures used at the time was sparse, leading to suspicions that the sites reflected convenience factors more than anything else (that is, ownership and site-specific knowledge). After a petition was delivered to the Prime Minister from the local residents of Billingham, and ICI withdrew their support, the site was abandoned and a decision on Elstow deferred pending a second site search. The choice of Billingham was a major mistake. The site is in the middle of a large urban area and obviously wholly unsuitable from the point of public and political acceptability, regardless of assurances of its safety. Failure at Billingham should have taught NIREX at least one vital lesson; the importance of a remote siting policy.

In a 'second attempt' site search, NIREX appeared more open about their methods. There seem to be three approximate stages in the site-selection process: survey, preliminary site identification, and site confirmation. These three stages are based around the International Atomic Energy Association (IAEA) 1983 guidelines but, like the DoE's (1985) guidelines, provide no definitive numeric siting criteria and were again mainly statements of principle. In the survey stage NIREX carried out simple 'pen and paper' sieve-mapping on four siting factors: geology, population distribution, accessibility and conservation issues (NIREX 1985 and McEwen and Balch 1987). The technique used appears no different to those developed by McHarg in the USA during the late 1960s (McHarg 1969). This traditional manual site identification process can only cope with very generalized map data and so is not able to perform a detailed and comprehensive site search.

An announcement in 1986 listed land adjacent to Bradwell nuclear power station, a disused airfield in Lincolnshire, and land formerly earmarked for an oil-fired power-station on the Humber Estuary as alternatives to be compared alongside Elstow (NIREX 1986). All four sites were owned either by the CEGB or the Ministry of Defence (MoD). Details of the second 'preliminary site identification' stage which identified these sites are poorly documented. Available literature does show how individual sites, mostly those within the areas identified in stage 1 (although not exclusively), were assessed for their suitability according to: site size and configuration; geology; planning issues; and ownership and availability (for example, McEwen and Balch 1987). However, precisely how these specific sites were pin-pointed from the hundreds of potentially suitable sites which must exist within the stage 1 areas is not clear, but favourable ownership and land availability appears to have again been major factors. The simple fact that all four selected sites identified in this way were owned either by the CEGB or MoD was enough to arouse public suspicions that the whole process had been 'gerrymandered' to prove the suitability of one or two preferred locations. Seemingly very limited regard was given to their true qualities as disposal sites and the whole search was based on only partial knowledge and without detailed consideration of the full range of alternatives.

This second attempt to find a waste site was abandoned in 1987 after detailed site evaluations revealed a supposed narrowing cost differential between deep and shallow land disposal and the uneconomic restriction of the sites to handle only LLW. A political factor may also have been involved; it was only a few months to a general election.

By the late 1980s research was being directed towards finding a deep disposal site to accept both LLW and ILW in one facility. Three disposal techniques have been proposed. These are: a land-based repository; a sub-seabed repository accessed from the coast; and a sub-seabed repository accessed from an offshore rig or artificial island (NIREX 1987). In the latter, it was envisaged that waste containers would either be deposited in caverns similar to the land-based and coastal designs or in deep boreholes. This option would utilize technology developed by the offshore oil and gas industry. NIREX have now narrowed the number of candidate deep disposal sites down to just two or three. These are sites at BNFL's Sellafield reprocessing plant, the UKAEA's Dounreay Atomic Energy Research Establishment, and possibly a further site near Dounreay in the Caithness granites, once investigated during the HLW drilling programme in the late 1970s and early 1980s. From this it would appear that the offshore option has been shelved. In view of this, the research presented here concentrates on the land-based and coastal disposal options.

It is still maintained by NIREX that initial site selection surveys should not be definitive and that a search for the 'best possible' site(s) would be impractical (McEwen and Balch 1987). Such a view is hard to believe given the recent advances in GIS technology that are detailed in the next two sections. It is apparent, however, that NIREX are employing the same selection process as used on the previous two failed site searches. They remain set on the 'engineering fix' with an overreliance on deep disposal for LLW and ILW alike as a 'Rolls Royce' solution to the disposal problem. However, the current short-list is clearly more pragmatic than previous attempts and seems, for the first time, to incorporate remote siting (especially Caithness) as a means of providing additional safety margins as argued by Openshaw (1986) for power-station siting. In addition, these sites appear adjacent to existing nuclear installations and so may take advantage of local support for the nuclear industry. This proximity may, of course, also be a major flaw. Should there ever be a major nuclear accident at the neighbouring nuclear site, then that accident could conceivably also disable the waste repository at a time when it might be most desperately needed.

Nevertheless, justification of the final chosen site will require a more rigorous siting evaluation than in previous UK nuclear inquiries as engineering arguments are no longer as convincing to the public as they were fifteen to twenty years ago. Indeed, from a narrow scientific point of view, the evidence that will be amassed to demonstrate that a repository is safe can never be convincing because the case can never be proven nor can any engineering design be tested under conditions equivalent to its operational life. It is argued, therefore, that it is worth while trying to change the emphasis from a purely untestable technical solution to a techno-geographical solution in order to find publicly-, politically- and environmentally-acceptable sites, whilst still retaining desired safety levels. The reasons for this include:

1 There is little benefit from the development for the local population.
2 The current plan is for only one facility receiving all the nation's LLW and ILW.
3 The national interest argument is overwhelming.
4 Whatever site(s) is selected there will be much public concern and opposition.
5 There can be no good reason for NIREX not wanting to find the 'best' compromise site and, having found it, prove its superiority.
6 The technology now exists to enable automated computer searches of *all* candidate sites to be performed in a broadly-based and locationally-comprehensive screening process.

In short, NIREX need to take the initial site-selection process more seriously. It is not merely a matter of trying to prove that site *X* is suitable, having stumbled on it by a circuitous route, but rather being able to demonstrate that site *X* was a reasonable choice in the first place. Previous nuclear public planning inquiries have concentrated on the former. A future radwaste dump inquiry may well not repeat this mistake.

OPPORTUNITIES FOR GEOGRAPHIC INFORMATION SYSTEMS

Commonsense suggests that if NIREX wants to convince people that they are doing their best to find an optimum location with the highest standards of safety and acceptability, then they need to switch from being happy with feasible sites to a search for 'best' sites, and then visibly demonstrate their near-optimality in respect to key criteria. Only this will enable NIREX to present the strongest possible case in justification for the development at any given location. To do this NIREX needs to adopt the best available technology to both make and validate their siting decisions.

The last decade has seen massive developments in computing hardware and the introduction of practical Geographic Information Systems. An increasing number of proprietory systems are now available, for example, Arc/Info and Spans. Indeed, so important is this field that the DoE commissioned a Committee of Enquiry on 'Handling Geographic Information' headed by Lord Chorley to investigate the potential of spatial data and GIS. The committee recommended that effort needs to be made to expand the awareness of potential users of GIS, particularly senior managers, so that its full potential might be realized. In light of the Chorley Report (1987), the DoE should now revise its siting guidelines to include the new opportunities for more rigorous analysis and evaluation made possible by GIS. NIREX should also fully embrace this technology in their site search and justification of siting decisions.

GIS offers a number of advantages over the more traditional manual approaches to site search and evaluation that NIREX have used hitherto:

1. It is fast, efficient and accurate.
2. The whole of the UK land surface can be searched and explored for suitable sites.
3. It is particularly good at handling a mixture of physical, socio-economic, environmental, and policy data sources.
4. It provides a basis for justifying the eventual siting decision which would be considerably stronger than that provided by more *ad hoc* methods.
5. In the eventual public inquiry, it is clearly an advantage to have an information system that can be used to demonstrate and investigate the strength of the case that is made.

To illustrate the advantages of the GIS approach, a demonstration system has been developed from digital map data using Arc/Info GIS software (Carver 1988). The objective of such a demonstration is threefold: to illustrate the capabilities of a GIS in this context; to show the existence of the necessary hardware, software and digital data; and to highlight the potential of GIS as an aid to developing a better understanding of the radwaste siting problem.

DATA SOURCES AND INPUT

A credible site search procedure can readily be developed using GIS. A number of digital map data sources are needed: accessibility data (based on road and rail networks), population data, geological details and conservation areas. The location of transport networks were taken from the Ordnance Survey 1:625,000 digital data base. These included rail and road networks as seen on Ordnance Survey 1:625,000 maps. Topographical information of interest, namely the UK coastline, was also taken from the Ordnance Survey 1:625,000 data. Information on population distribution was taken from the UK census. Population densities have been derived from these data for one kilometre grid squares for the whole of the UK which were used to define areas of high population density. Data from the 1971 1 kilometre grid square census files and estimated 1981 1 kilometre data were used. Built-up area boundaries, a surrogate variable for highly-populated areas, were again taken from Ordnance Survey 1:625,000 data.

The other data used were specially digitized. Geological areas were digitized direct from maps (produced by NIREX by the British Geological Survey (BGS)) which identify those areas most likely to have favourable hydrogeological environments at depth for a land-based or coastal repository (Chapman *et al.* 1986). Conservation areas were digitized from Countryside Commission maps. Small conservation sites too small to be shown on large-scale maps were first identified using the BBC Domesday Advanced Interactive Video (AIV), then input manually as point references and then buffered at small (1 km) radii. Buffering is an operation implemented in the GIS software. It involves placing a mapped feature (point, line or polygon) inside a zone of constant specified width.

POLYGON OVERLAYS AS A MEANS OF DEFINING SUITABLE SITES

Each of the three alternative repository designs by their very nature define their own areas of search. A land-based repository can only be located within the British coastline, whilst a sub-seabed repository accessed from a rig-type structure must be located in an offshore position. A coastal repository must be located within a coastal area, with its shaft entrance on the coast and its disposal vaults some 2 to 5 kilometres from the shore. These three different areas of search were defined using the UK coastline and International Meridian: the coastline defining the search area for a land-based repository; a 5 km coastal buffer zone for a coastal repository; and areas more than 5 km from the shore and within the International Meridian for an offshore site.

These three search areas provide the basis for two separate site searches, one combined search for land-based and coastal sites, and one for offshore sites. Areas of suitable hydrogeological environments in each of the three search areas then provide a starting point from which all potential sites in the UK that satisfy various combinations of criteria can be identified. This has been achieved by placing digital map coverages through an automated sieve-mapping process utilizing the polygon overlay functions within Arc/Info to identify those areas which simultaneously satisfy all the siting criteria. Since IAEA and DoE siting guidelines contain no definitive numeric siting criteria, arbitrary criteria were assigned to the data to implement the site search. However, when possible, numeric criteria were chosen which reflect those used by NIREX in previous site searches.

The GIS procedure for the identification of areas containing sites suitable for land-based and coastal facilities involved a number of stages. These are:

1 Define a broad search region as being the whole of Britain's land surface and up to 5 kilometres from the coastline.
2 Define areas of suitable geology (Figure 7.1).
3 Remove all locations with a 1 kilometre population density greater than 490 persons (Figure 7.2).
4 Exclude areas not within 3 kilometres of a railway or motorway and 1.5 kilometres of a primary road (Figure 7.3).
5 Exclude areas of high conservation status (Figure 7.4).

A population density of 4.9 persons per hectare has been used hitherto by NIREX. This was derived by them from the Nuclear Installation Inspectorate's (NII's) relaxed nuclear power-station siting guidelines of 10,000 people within a 5-mile radius (McEwen and Balch 1987), although the current power-station siting criteria cover a much larger distance. NIREX argue that the safety of a repository has to be so high as to make attention to population distribution unnecessary. Nevertheless, it must be considered expedient to promote a remote siting policy. NIREX do this with reference to the above population limit, which effectively ensures that the minimum area of the country is excluded. A proper remote siting policy would consider areas with only very low population densities, and perhaps more importantly, areas of low population accessibility. It can be argued that only by considering such really remote areas can the effects of the 'Not In My Back Yard' (NIMBY) syndrome be minimized. Site searches have been carried out which utilize more restrictive population criteria (Openshaw *et al.* 1989, and Carver 1990).

Rather than place too much emphasis on 'strategic' accessibility (that is, Weberian type isodopanes describing accessibility to producer sites) which makes strong assumptions about current operational transport costs, packaging volumes and so on, local accessibility is stressed by buffering road and rail networks. The buffer widths of 3 and 1.5 kilometres are arbitrary, but appear reasonable in the context of the current site-search process. Choosing areas in close proximity to transport routes both minimizes waste handling (that is, transferral of waste between different modes of transport), thereby reducing operator dosage and accident risk, and also minimizes the need to construct new transport links to serve the disposal facility.

Finally, the definition of areas of high conservation status is also arbitrary and in reality may not be the same as used here. Nevertheless, the list should include areas defined as National Parks, Areas of Outstanding Natural Beauty, Heritage Coasts, National Scenic Areas, Environmentally Sensitive Areas and Regional Parks by the Countryside Commission. This eliminates most of the more contentious and scenically and/or ecologically valuable areas from consideration.

It is useful here to stress that all the siting criteria used are flexible and can be changed as desired to meet individual user requirements.

'WHAT IF?' MODELLING

Throughout this process of polygon overlays, the speed and flexibility of a GIS allows quick comparative re-evaluation or 'What if?' modelling to be carried out on the selected siting criteria. This allows the developer and decision-maker to test 'what if this condition or that criterion is changed?' and see the effect on the final map. Such an exercise has been performed on the population data. Population density criteria were tightened up by applying the NII's

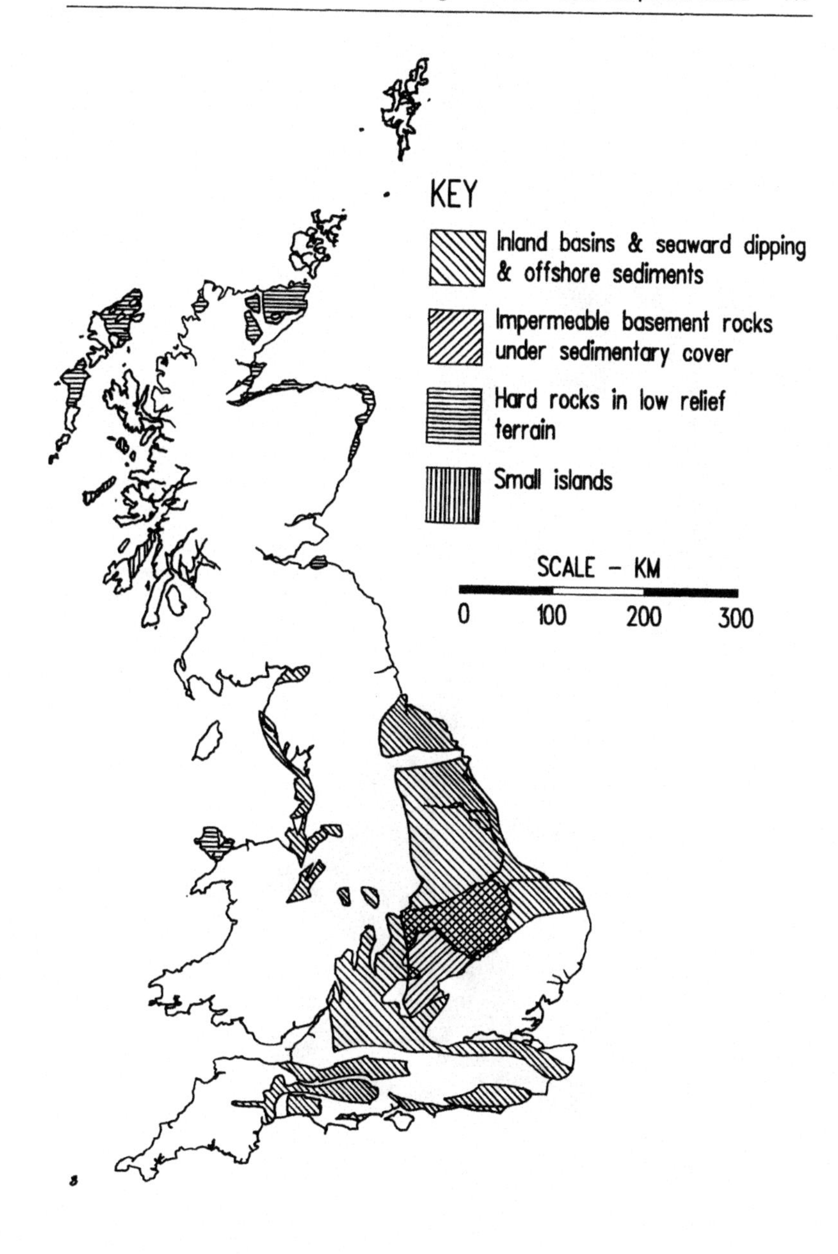

Figure 7.1 Deep hydrological environments considered suitable for low- and intermediate-level radioactive waste disposal

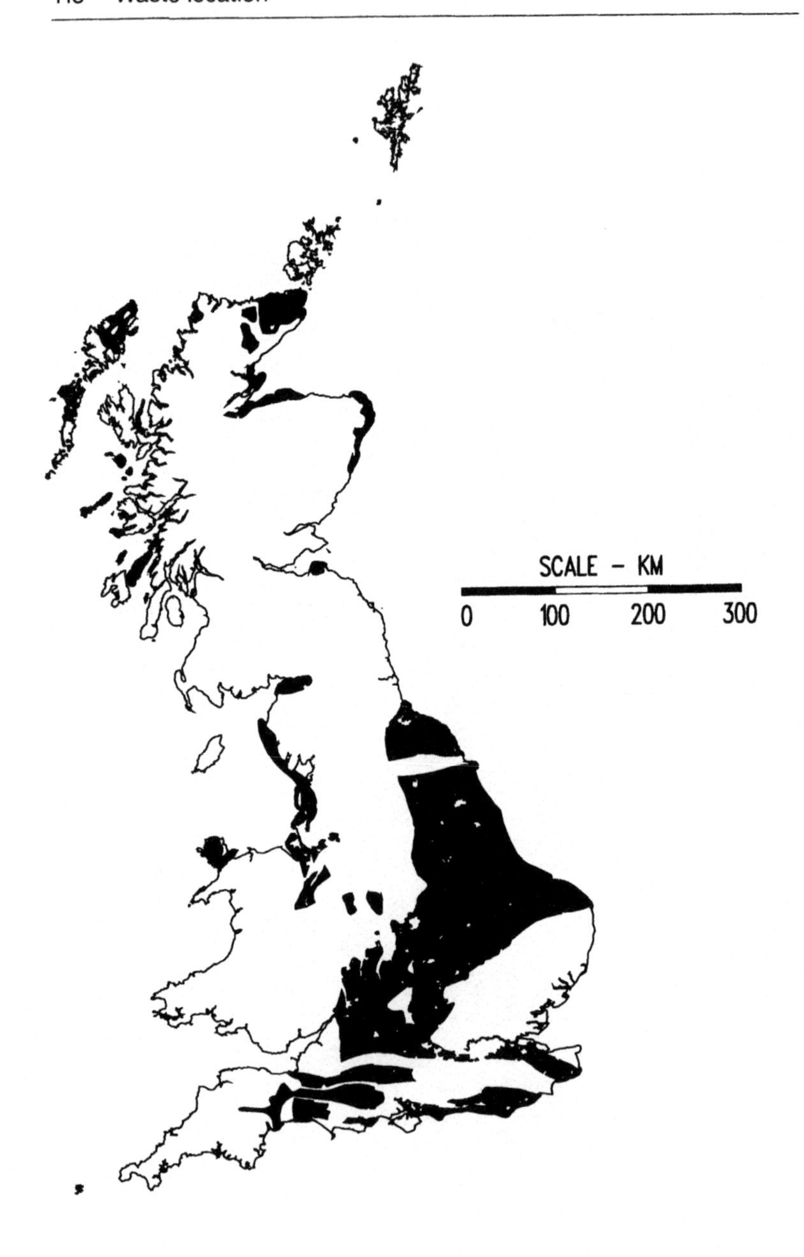

Figure 7.2 Feasible areas after removal of populated grid squares with density greater than 490 persons per square kilometre

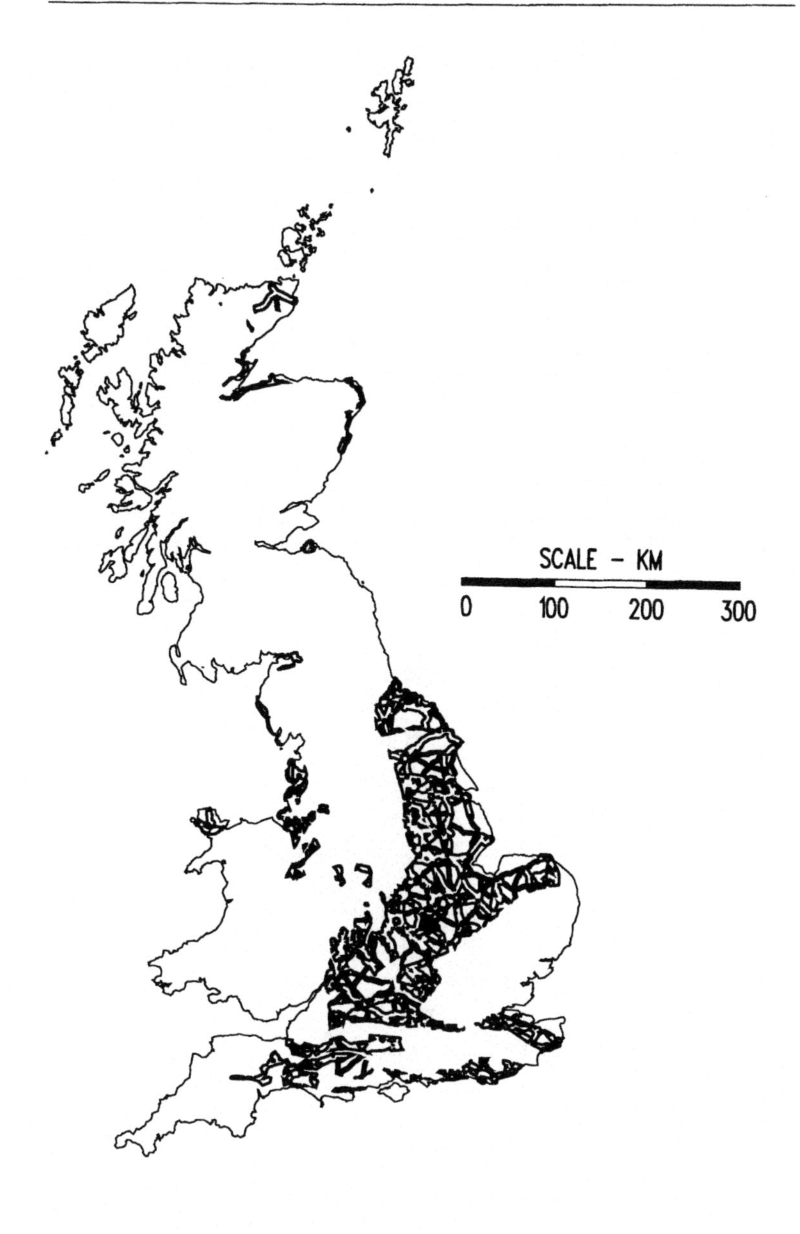

Figure 7.3 Feasible areas after removal of areas outside access corridors (3 km rail and motorway buffer and 1.5 km primary route buffer)

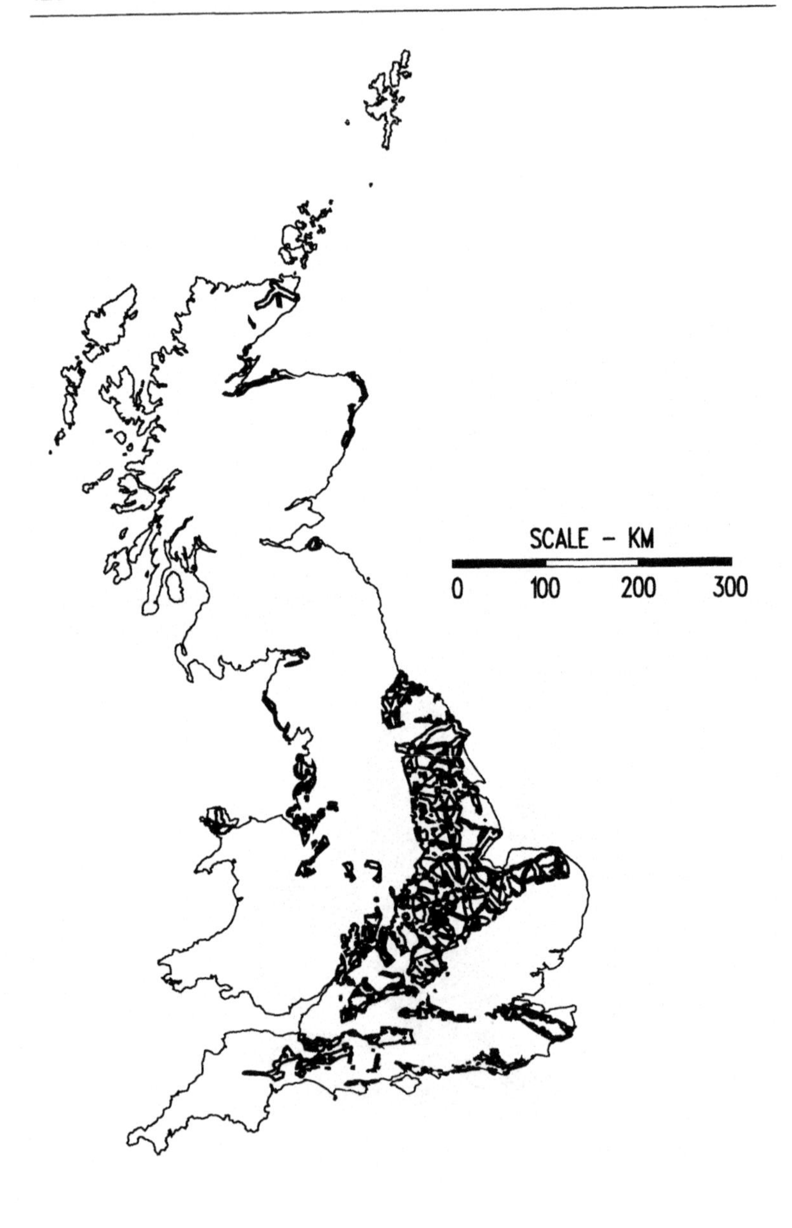

Figure 7.4 Final feasible areas after removal of conservation areas, using criterion for Figures 7.2 and 7.3

original unrelaxed power-station siting guidelines of 100,000 people within a 5-mile radius, which yields a population density of 490 persons per kilometre square. It is important to note that NIREX's interpretation of the nuclear power-station siting guidelines is a corruption of the NII's original definition. This definition sets a limit to the total population within a 5-mile radius and as such implies a degree of remoteness. NIREX's recalculation of these figures on a 1 kilometre grid square basis clearly does not. As a result small areas, apparently within populated areas, may remain at this stage. This is especially true in populated coastal locations where unpopulated inshore areas of sea adjoin coastal towns. Such areas would be excluded later in the site search by factors describing site remoteness or population accessibility.

Conditions applying to the assumed mode of transport have also been altered. Here it is assumed that an all rail transport policy is to be pursued, so that only those areas within 3 kilometres of a railway line are retained. Both are relatively major alterations in siting criteria, but they nicely illustrate the effect changes in site selection criteria can have on the final map (Figures 7.5 and 7.6).

This exploratory process is a vital ingredient in demonstrating the validity of a particular site. A 'good' location should be fairly robust and an ability to experiment with the criteria is an extremely useful means of testing this. The national interest is best served if there can be some broadly-based agreement about where the best sites are located. This cannot be achieved by GIS alone. However, GIS does provide a means of debate that could, when operated by reasonable people, assist in the process of gaining a widely-based consensus of opinion. One problem in the UK is the absence of any framework within which this 'debate' can occur.

MULTI-CRITERIA EVALUATION

The mapped results in Figures 7.4, 7.5 and 7.6 are no more than areas which are feasible in the context of the siting criteria used in the GIS polygon overlay process. All these locations are not of equal suitability. A further stage in the analysis could be to apply multi-criteria evaluation (MCE) methods to identify optimum sites for the decision-maker. A list of potentially suitable methods is given by Nijkamp (1980) and Voogd (1983), for example, hierarchical optimization. Here we do no more than outline the methodology of how such procedures might be applied.

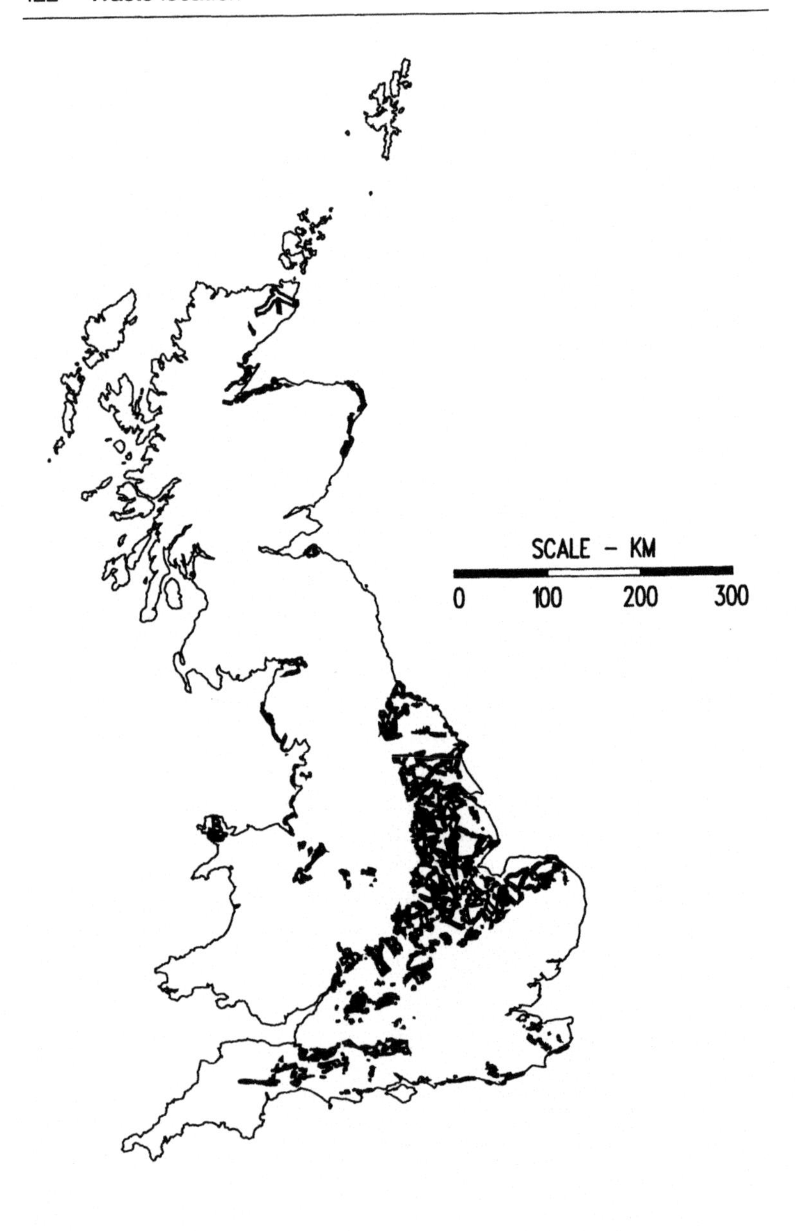

Figure 7.5 Feasible areas after removal of populated grid squares with density greater than 490 persons per square mile

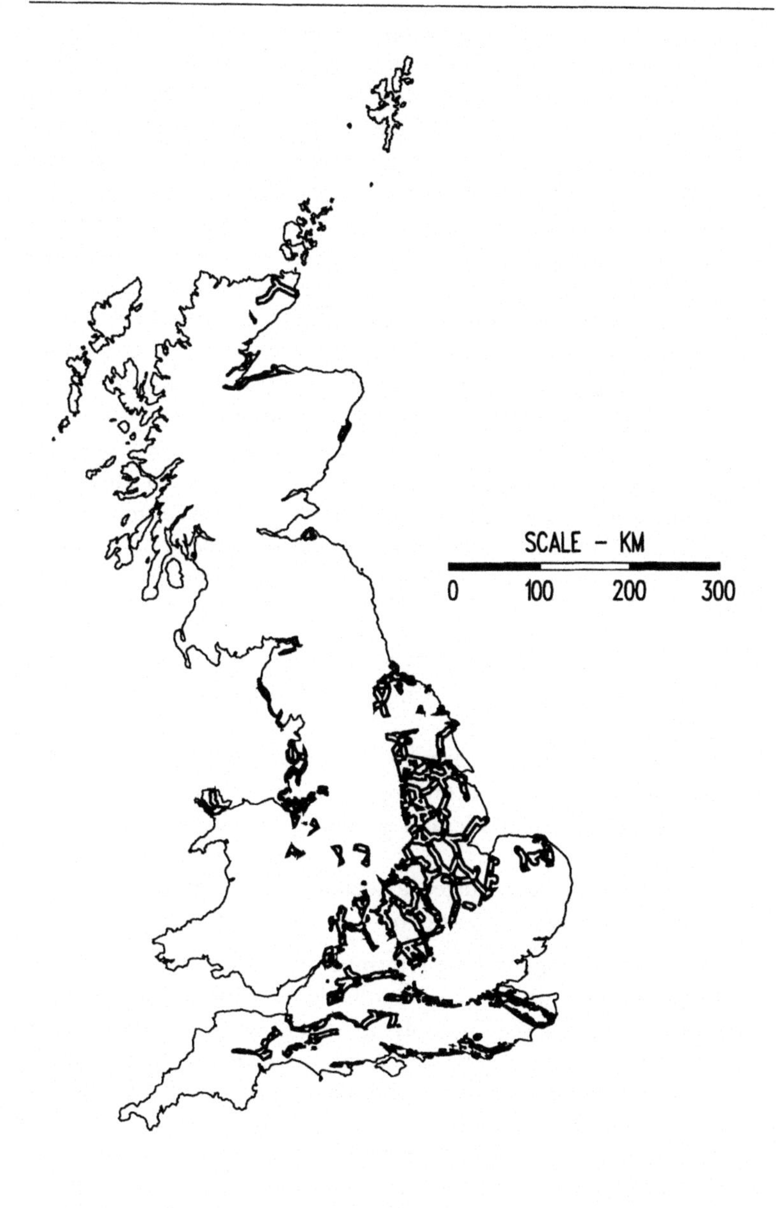

Figure 7.6 Feasible areas after removal of areas outside rail-only access corridors (3 km rail buffer)

First, the surviving area polygon coverages from the GIS overlays are reformatted into 2 by 2 kilometre grid squares or rasters. Each raster is then regarded as a separate 'site'-sized land parcel. This has been achieved by combining the surviving area polygons along with a 2 by 2 kilometre grid, using the grid square centroids in a 'point in polygon' exercise. Centroids falling inside surviving area polygons are used to define the 'site' rasters and are stored as grid references in a single variable file (SVF). Data from relevant national data bases relating to secondary or 'soft' siting considerations is assigned to each of the rasters contained in the SVF. Data include: dominant land use; absolute population density; accessibility to population centres; strategic accessibility; quality of local transport links; constructability; resource sterilization, and so on. Computer-implemented MCE techniques are then used to apply predetermined weightings to the data and simultaneously compare each site against all the other alternatives. From this a short-list of 'best available sites' could be produced. The advantages obtained by using MCE within a GIS framework for short-listing candidate sites from feasible areas include:

1 It allows complex trade-offs to be performed on multiple and varied siting factors by single or multiple sets of decision-makers.
2 It allows value judgements to be incorporated into the analysis by weighting factors according to their perceived importance.
3 It provides a systematic and semi-objective framework of analysis.
4 It has all the advantages associated with the use of a GIS toolbox data-base construction.

However, MCE cannot be performed in a vacuum. The results would be conditional on the trade-off decisions made, ideally within an explicit MCE or else invisibly, to select the short-list within the set of feasible locations. The nature of the data and the techniques used to draw up the short-list of sites dictates that a certain degree of 'fuzziness' will occur between individual sites and as such it is unlikely that any one site could be identified as 'number one'. At best, groupings of say, the top ten, the next ten, and so on down the ranked list of sites, could be achieved. Further short-lists should be based on a selection from each of these rough groupings. The degree of diversity, thus retained, should allow for some choice between the final three or four sites. This is not the same as picking sites purely to demonstrate the superiority of an *a priori* defined and preferred location. This process should be an honest one and the public inquiry must be in a position to require that this is proven.

RESULTS AND CONCLUSIONS

The results from GIS polygon overlays of this exercise can be seen in the form of a number of final maps describing surviving areas which satisfy all the specified site search criteria (Figures 7.4, 7.5 and 7.6). Results from the MCE stage consist of a short-list of 2 by 2 kilometre 'sites' with near-optimum qualities as a disposal site for each of the repository designs under consideration. This approach makes the best use of vector and raster data with the Arc/Info GIS in such a way whereby each is used for those problems it is best suited for. Vector data is the most appropriate way of dealing with large polygons, line features and point references as found in stage one. The raster data becomes a very useful way of handling more refined site evaluation in stage two, where there are no longer any deterministically applicable rules.

It must be stressed that the site searches described here are not intended to be a replication of the NIREX siting process, but are purely to demonstrate some of the capabilities and advantages of GIS in this context. There are parallels with what seems to be happening at present with the important exception that the comprehensiveness of the site search procedure, the speed and efficiency of operation is dramatically improved by the utilization of GIS. Consider manual sieve-mapping; the level of detail obtainable by this method must be considered at best inadequate, given the importance of the task in hand. This lack of accuracy is especially apparent when considering geographically complex factors like population. For example, NIREX used districts and local authorities as the areal units for mapping population density, resulting in an over-generalized national picture of population distribution. Obviously nobody can be expected to carry out manual sieve-mapping with the 1 kilometre population data used here, but using GIS the task is very simple and the far greater resolution provided means greater confidence in decisions made as a result.

Technically the combined approach is superior to traditional manual techniques in terms of accuracy, speed and efficiency. The most important advantage that GIS has over the traditional methods, however, is its flexibility and the highly communicative nature of its output. The ability to ask 'What if?' type questions in a public consultation environment would provide both sides with valuable information needed to both develop and defend their arguments. Each side can stress those siting factors and set criteria to identify sites which are both safe and acceptable to them. The use of GIS requires that there is no monopoly on the technology and that all sides have access to the same GIS and sufficient knowledge of the technology so as to be able to articulate their dabate through it. This would be possible using customized interfaces and independent operators.

Inevitably different parties will stress different criteria. With the criteria and their relative weights identified, an independent expert operator could identify all the compromise sites using a GIS system and data base. High resolution colour graphics could be used to provide a clear visualization of the sequence of operations carried out in identifying compromise sites and show all alternatives for further discussion. This could, if implemented properly, represent a considerable improvement on traditional public consultation exercises and provide a powerful tool for rational political and scientific debate. It should then be up to the government to ensure that unbiased ground surveys are carried out to assess the ultimate suitability of the chosen compromise sites.

Failure to adopt GIS in the future as a means of implementing a comprehensive nationwide site search may well result in NIREX failing to hold true public consultations and being unable to demonstrate to the public and scientific community that all suitable alternatives have been identified and considered. Until GIS technology is adopted, whatever sites NIREX may select will be scientifically suspect, open to challenge and based on a partial and potentially-biased examination of the British land and sea space. NIREX would be well advised to consider the use of GIS as a means of ensuring greater understanding of the site search process, of gaining greater public acceptibility for their proposals and minimizing the risk of yet another failed site search.

REFERENCES

Carver, S.J. (1988) 'The role of GIS in radioactive waste disposal', paper given at Environmental Systems Research Institute (ESRI) User Conference, Freising University, West Germany, October.

Carver, S.J. (1990) 'Application of Geographic Information Systems to siting radioactive waste disposal facilities' unpublished PhD thesis, University of Newcastle upon Tyne.

Chapman, N.A., McEwen, T.J. and Beale, H. (1986) 'Geological environments for deep disposal of intermediate level wastes in the UK', paper given at IAEA International symposium on *The Siting, Design and Construction of Underground Repositories for Radioactive Wastes*, Hannover, March, IAEA Report no. IAEA-SM-289/37.

Chorley, R. (1987) *Handling Geographic Information: report of the Committee of Enquiry*, London: HMSO.

Department of the Environment (1983) 'Disposal facilities on land for low and intermediate level radioactive waste: draft principles for the protection of the human environment', London: HMSO.

Department of the Environment (1985) 'Disposal facilities on land for low and intermediate level radioactive waste: principles for the protection of the human environment', London: HMSO.

House of Commons Select Committee on the Environment (1986) 'First report from the Environment Committee on radioactive waste', London: HMSO.

International Atomic Energy Agency (IAEA) (1983) 'Disposal of low and intermediate level solid radioactive wastes in rock cavities', IAEA Safety Series no. 59, Vienna.

Layfield, F. (1987) *Sizewell B Public Inquiry*, Department of Energy, London: HMSO.

McEwen, T.J. and Balch, C. (1987) 'Geological and environmental constraints on the disposal of low level wastes in the UK', *Planning and Engineering Geology*, London: Geological Society.

McHarg, I.L. (1969) *Design with Nature*, New York: Natural History Press.

Nijkamp, P (1980) *Environmental Policy Analysis: Operational Methods and Models*, Chichester: Wiley.

NIREX (1985) *Plaintalk*, free newspaper.

Nuclear Industry Radioactive Waste Executive. (NIREX) (1986) 'NIREX site announcements: information from NIREX', UK NIREX Ltd, February, Harwell.

Nuclear Industry Radioactive Waste Executive (NIREX) (1987) 'The Way Forward', UK NIREX Ltd, November, Harwell.

Openshaw, S. (1986) *Britain's Nuclear Power: Safety and Siting*, London: Routledge.

Openshaw, S., Carver, S.J. and Fernie, J. (1989) *Britain's Nuclear Waste: Safety and Siting*, London: Belhaven Press.

Parker, R.J. (1978) 'The Windscale Inquiry', London: HMSO.

Radioactive Waste Management Advisory Committee (1988) *Ninth Annual Report*, London: HMSO.

Royal Commission on Environmental Pollution (1976) *Sixth Report: Nuclear Power and the Environment (The Flowers Report)*, Cmnd 6618. London: HMSO.

Voogd, H (1983) *Multi-criteria Evaluation for Urban and Regional Planning*, London: Pion.

Chapter 8

Heavy metals in soils and diabetes in Tyneside

Simon Raybould, Y. Crow and K.G.M.M. Alberti

INTRODUCTION

Environmental epidemiology is a contentious issue. It is remarkably easy to generate huge headlines and press speculation, often citing 'causes' and 'proof' when all that is achievable is an association. This notwithstanding, its value should not be underestimated, in that it provides, at the very least, a background for more clinically-orientated studies or more often a framework in which they can be undertaken; certain lines of exploration can be 'ruled out' as it were, before expensive clinical trials are carried out.

Diabetes mellitus is a disorder in which blood glucose levels are high due either to lack of production of insulin, or inadequate action of insulin, or a mixture of the two. It exists in two main forms: insulin-dependent diabetes mellitus (IDDM) and non-insulin-dependent diabetes mellitus (NIDDM) (WHO 1985). The latter is extremely common world-wide and is related to age and life-style, including lack of exercise and being overweight. IDDM is an auto-immune disorder, occurs in young people, results in death if untreated, and is concentrated in northern areas of the world, being most common in Europeans. There is a strong inherited component, but environmental factors are also thought to be important. Viral infections may play a role but there must be other, as yet unknown, environmental factors. A relationship with social deprivation has been suggested recently in the northern region of the UK (Crow, Alberti and Parkin 1991).

This chapter is taken from work which arose from the coming together of several factors, largely relating to data availability. Such a state of affairs is highly fortuitous, in the light of the current concerns relating to the 'green debate'; there is burgeoning interest

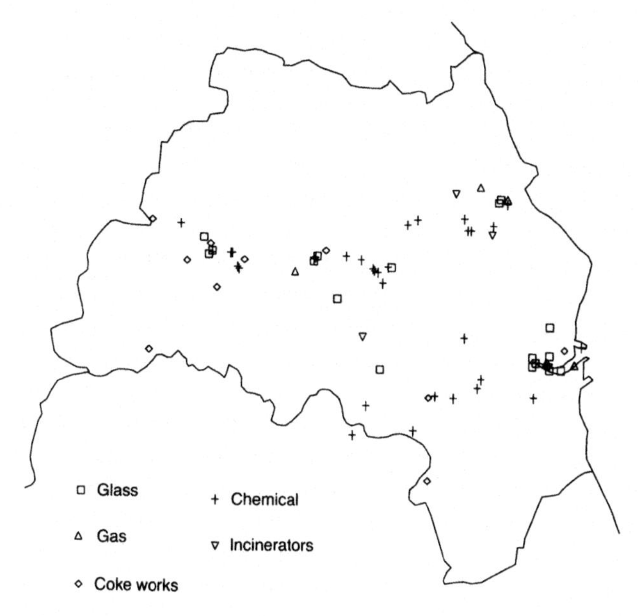

Figure 8.1 Industrial activity in Tyneside

in all issues relating to industrial development as a whole. The so-called 'heavy metals' (however defined) are a particular issue of concern – perhaps because of their association in the public's mind with 'all things atomic' – but also more seriously as a matter of some debate in academic medical literature. Particularly of concern are the factors of heavy metal origin: they are produced, for example, as a by-product of the urban waste disposal systems of many parts of the country (both industrial and domestic) by way of the combustion processes in incinerators. To make matters worse, *municipal* incinerators, at least, are often situated in a residential environment.

The aim of this chapter is straightforward: to undertake an evaluation of the level of spatial association between the locations of diabetic subjects in Tyneside and the amounts of various heavy metals present in the soil – irrespective of whether that level is natural or the results of human activities. (Tyneside is a noted industrial area of the country with a plethora of potentially polluting activities scattered across its surface. Figure 8.1 gives a simplified indication of some of them.)

Conversely, the practical problems of such an investigation are, unfortunately, as complex as the questions are simple. Data have to be integrated which relate to the location of the cases, the pattern of the background (or 'at-risk') population and the actual level of heavy metals: no attempt is made to investigate the issue of behavioural and physiological pathways, as this is an essentially exploratory work, and the assumption is therefore made that the effects of such pathways are spatially uniform. Furthermore, conventional spatial methodologies are not appropriate here because of the relative sparcity of cases, which exacerbates the already potentially serious problem of the modifiable areal unit problem (MAUP) (Openshaw 1985). These considerations lead to the adaptation of a Monte Carlo approach, which is essentially a simple modelling methodology. A conceptual model of the full hypothesis is set up and the results of running an operationalized model, which simulates this, is compared to the observed pattern of events. In this instance diabetes cases were compared to the locations within Tyneside which were chosen purely on the basis of the distribution of the at-risk population, with no reference to any supposed environmental 'causes' of diabetes, which were to be investigated.

Also for these reasons, a great deal of attention is paid to methodological issues in this chapter.

THE DATA

The diabetes data

All subjects with diabetes diagnosed and treated with insulin before their 16th birthday and resident within the northern region at diagnosis were sought for the years 1977 to 1986 inclusive. South Cumbria District Health Authority was excluded because of cross-boundary flow of patients. Names were obtained from the Regional Health Authority computer print-out of all patients coded for diabetes mellitus. All newly-diagnosed patients under 16 years old would be admitted to hospital. All case notes were then checked for eligibility for inclusion and date of birth, date of diagnosis and the postcode at the time of diagnosis recorded. Validation was carried out by: (1) questionnaires to consultant paediatricians, (2) clinic registers, (3) community paediatric registers. A total of 894 individuals were identified. It was estimated that the primary data set was 95 per cent complete, and validation sources identified 54 per cent of the total cases, well above the 30–40 per cent usually found by this method.

The key to *spatial* analysis of these data is the successful completion of the postcode field. The unit's postcode consists of up to seven letters and digits, with a space separating the last number-letter-letter sequence from the rest. They are usually streets, or parts of streets with an average of fifteen households to each postcode. By use of the postcode address file (PAF) it is possible to ascribe to each case a grid reference, with a spatial resolution of 100 metres, based upon its postcode. This form of spatial referencing allows comparisons to be made with other data-sets by using a common means of identifying locations. In fact 4.8 per cent of cases had to be excluded from the analysis at this stage due to a failure to convert postcodes to grid references. The reasons for failure are given in Table 8.1.

There is no reason to suppose that there is a spatial bias in the 'missing' or 'incomplete or incorrect' postcode fields. It is possible however that those cases which were lost because they were not on the PAF could be spatially clustered (a housing estate, for example, that is not included on the PAF for whatever reason). As such cases are such a small percentage of the total data set, however, the problem was not treated as significant enough to warrant the extensive manual work necessary to locate the cases.

The population data

The second unusual (in fact unique) data set which was used in the analysis was the Tyne and Wear Joint Information System (or JIS). This is a country-wide data base maintained by the Research and Intelligence Unit of the residual body of the now-defunct Tyne and Wear County Council.

The JIS was started in 1975, and has a federal structure. Most of the data are therefore stored and updated at the level of the participating districts within the county – Newcastle, Gateshead, North and South Tyneside and Sunderland (including Washington). The JIS can be regarded as a computer-based data-set which holds one record per case for all rating hereditaments in the county. In terms of residential hereditaments, each case will therefore equate to a household. It provides an almost ideal basis for creating a sampling-frame for subsequent use.

The primary weakness of the JIS as a sampling-frame (the use to which it is put in this chapter) is related to its updating system. Each month 'change data' are gathered by routine reference to regular administrative and statutory information flows within eight or nine

Table 8.1 Diabetes data removal

Cause	*Percentage*
Missing postcode	3.1
Incomplete or incorrect postcode	1.2
Postcode missing from PAF	0.5
Total	4.8

departments in each of the county's districts. This is supplemented by the results of irregular inspections and surveys: additionally, departments already using the JIS gazetteer provide a source of feedback relating to apparent errors and omissions. The data provided by each of these sources are collated and checked by comparison with each other, and also against the existing situation, as depicted by the state of the gazetteer for the previous month. Only if apparent changes or errors are thus validated is a change made to the appropriate records throughout the system's data bases, including the base maps.

The results of this level of interactive checking, together with occasional *ad hoc* accuracy checks, are twofold. Beneficially, the data contained within the gazetteer can be relied upon to be almost totally accurate and comprehensive, but unfortunately there is no historical content to the data. In other words, once a change is made to a property, and subsequently therefore to the JIS gazetteer, its previous land-use code, for example, is lost. There is no systematic way of recovering such data.

The ultimate effect of this updating system for the use of the JIS gazetteer as a sampling-frame is that changes in land use, without associated physical changes, or changes to the physical structure of the conurbation itself, such as slum clearance/redevelopment and so on, are not allowed for in the frame; the methodology assumes that the state of the data at December 1985 (the time when the JIS staff generated the copy of the gazetteer) is perfectly representative of the situation over the study period, and that no changes at all have taken place. This is a somewhat tenuous assumption, as the North-East of England, as a region, is noted for the number of slum clearance projects undertaken in recent decades.

The acceptance of the assumption that no changes took place in the infrastructure of the county during the study period is obviously undesirable, but it is somewhat hard to avoid. In any case, the subsequent integration of the JIS with data from the 1981 Census of Population (which was necessary to provide information on the background (or 'at risk') population distribution) meant that such a dynamic component of the JIS data would have been only a fallacious advantage.

Setting aside the above problems, such integration of these data sets allows one to avoid the very severe problems of the MAUP which have typically plagued work such as this, although it did restrict the analysis to only the one county. This meant that only 361 cases were used in the final analysis.

General background data on the population of the area was provided by the 1981 Census, accessed at the Enumeration District (ED) level.

DATA INTEGRATION FOR SUBSEQUENT USE

Introduction

The method chosen for this analysis was Monte Carlo simulation. This is a slightly unorthodox technique believed to be simultaneously quite robust and sensitive (Hope 1968; Raybould 1987).

Prior to using such a method, however, a suitable sampling-frame must be identified. By assuming that it is not unreasonable to associate the postcodes of diabetics with diabetic people, it is possible to do this by integrating the JIS with the 1981 Census. This is obviously a compromise from the ideal 'locational unit' of the individual, but it is none the less heuristically acceptable. The ideal units would be, of course, the individuals of the population at risk. Unfortunately, individual data of this nature are not available in the UK, so surrogacy techniques must be adopted in an attempt to get as near this idealized situation as possible, in this case by weighting households by a measure of the at-risk population density in the surrounding area – in this case the ED.

In the succeeding sections of this discussion the process by which households were selected as part of a simulation data-set is outlined. The point of such procedures is to modify data for locations in such a way as to maximize the level to which they reflect actual variations of the at-risk population.

The integration of census and JIS data and the selection of cases

The actual details of the selection procedure are somewhat complex, but are of fundamental importance in establishing the validity of the technique. Figure 8.2 gives a conceptual indication of the procedures which were undertaken in order to generate a set of control points, and their associated dates. The concept was relatively simple, in that the probability of any given member of the at-risk population living in Tyneside over the study period being diabetic could be simply calculated as being equal to 'x', where:

x = number of cases/number of members of the 'at-risk' population,

and the 'at-risk' population was calculated from the 1981 Census. This rate was subsequently used in generating the controls from the sampling-frame. This was differentiated from the gazetteer itself by the addition of supplementary demographic data. These background data were obtained from the Census. They were integrated into the gazetteer by weighting the probability of any given individual being diabetic (x) by the probable number of members of the at-risk population resident at the one household. This probability was based upon the number of households in the same ED as the postcode in question which had a given population of the appropriate age. A household in the gazetteer with only one resident will have a probability of being selected in the Monte Carlo procedures of equal to x; a household with two residents will have a probability of equal to $2x$, and so on. This value is referred to as 'alpha'.

One source of error in the above procedures is illustrated in Figure 8.3. The actual source of the error lies in the shaded subset where *Pa intersects Pb*. The procedure outlined makes the implicit assumption that cases of diabetes are dependent upon each other in that the presence of one diabetic in a household precludes the possibility of another case occurring in the same household, as the same household cannot be selected more than once in any given simulation run (see later). Figure 8.3 shows the case where both residents of a two-person household have diabetes. The likelihood of such a case occurring can be calculated as:

$$Pab = 2x - x^2$$

where:

Pab = probability of both resident A and B being diabetic
x = probability of any given resident of the area being diabetic (that is, 'A' or 'B' being diabetic)

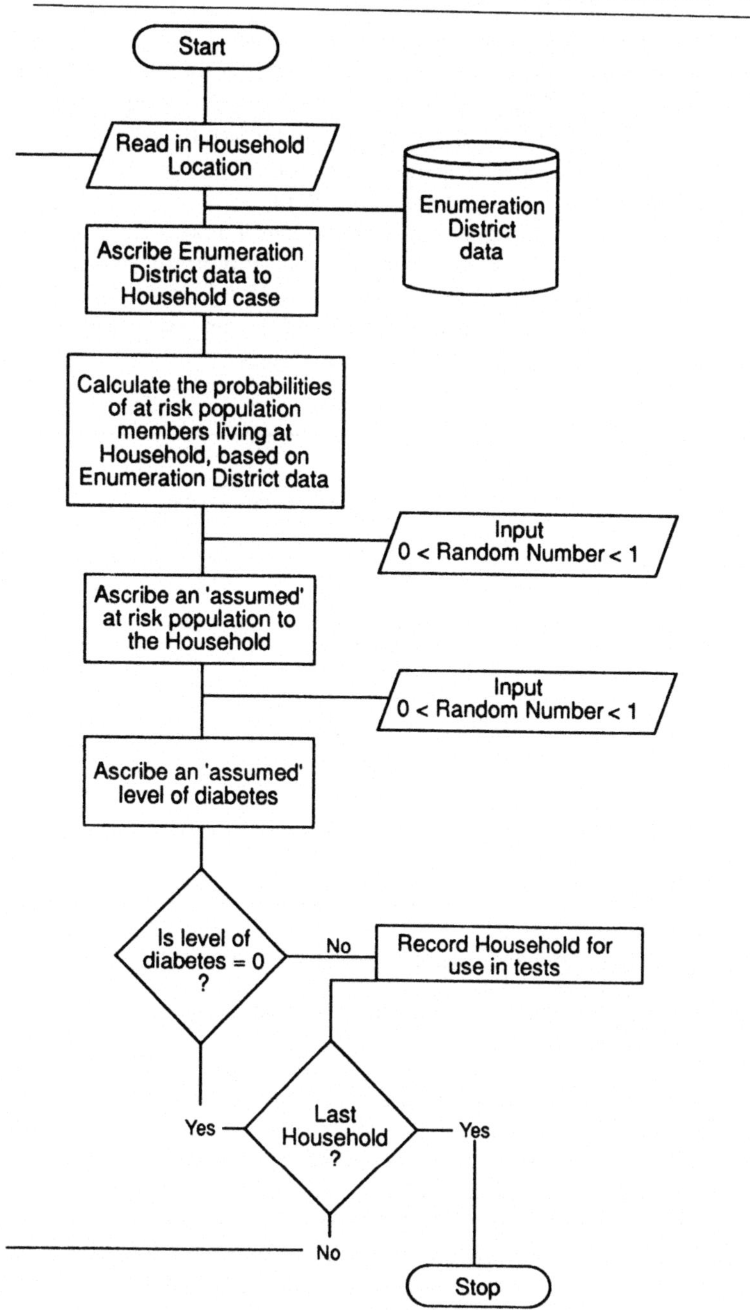

Figure 8.2 Flow diagram of integration

This could be a problem in that IDDM does occur periodically amongst siblings. Thus it has been estimated that there is a 5–10 per cent chance of a sibling of an IDDM patient developing diabetes, rising to 12–24 per cent if the non-diabetic sibling is genetically similar with regard to HLA (Human Leucocyte Antigens). However, there were very few such siblings in our data.

In practice only 2.74 per cent of cases shared a postcode with any other case: given further, that each postcode represents an aggregation of (on average) fifteen households, familiar clustering – resulting in several cases in one household – was not considered a significant problem from a spatial point of view.

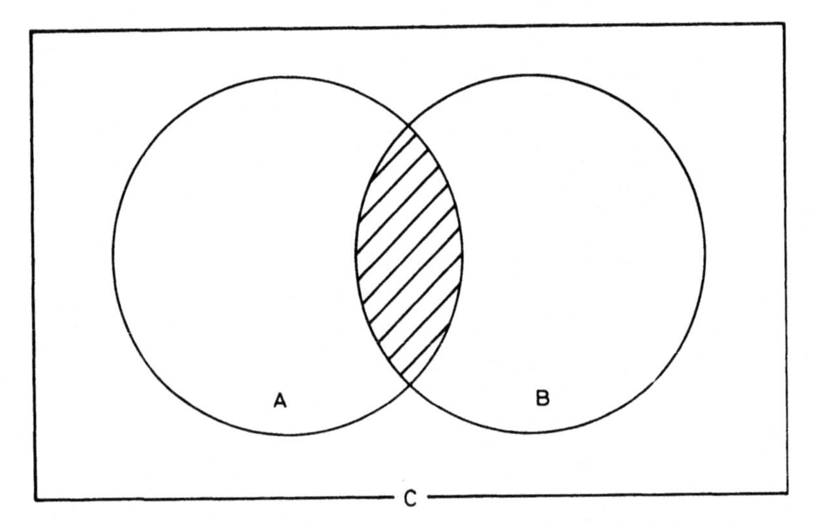

Figure 8.3 Diagrammatic description of an error source in the Monte Carlo procedure
Note: A + B = children in household. C = universal set of all children in household

Once an acceptable value for alpha has been defined for the household, a random number is generated, lying between 0 and 1.0. This number in turn is compared to the probability of the household in question have a diabetic case in residence (alpha), and the household is accepted into the control data-set or rejected on the basis of this (that is, if the random number is greater than alpha the household is rejected, otherwise its details are stored as part of the control data-set).

A summary of the approach represented by the use of Monte Carlo simulations just outlined could be regarded as the comparison of the actual distribution of diabetes with that of a hypothetical model. This model is operationalized in the form of the selection procedure for the control locations to be used in those comparisons. The assumptions of this model are shown in Table 8.2.

Table 8.2 Assumptions made in using Monto Carlo methods

Model assumption	*Manifestation of assumption in the selection procedure*
(1) Events are totally independent of time.	Controls are randomly ascribed onset from a uniform distribution.
(2) Events are related to the locations of children in the study area.	Controls are selected in a quasi-random fashion from households in the study area, weighted by at-risk population.
(3) Events are independent of all environmental stimulants, such as pollutant levels, etc.	Controls are chosen with no reference to any exogenous environmental factors.

DISCUSSION OF THE ASSUMPTIONS IMPLICIT IN THIS APPROACH

A brief discussion of several assumptions which are implicit in this approach is in order here.

It would obviously be ideal if data were available at an individual level which gave information on the distribution of numbers of the at-risk population. This is quite simply not possible at the moment in this country, as such data are not released by the Office of Population Censuses and Surveys (OPCS); the lowest level of aggregation is the ED. Up to a point it is reasonable to use the distribution of houses and flats, and so on (that is, the sampling-frame, based upon the gazetteer data) as a surrogate for the distribution of the population. If it was possible to assume that the occupation density for children remained constant within households throughout Tyne and Wear, this would be a sound assumption from which to work. Unfortunately, densities of occupation do vary between households. The best that can be achieved therefore, is a combination of the census and gazetteer data-sets in the manner outlined in the previous section to ensure that the gazetteer entries are 'sensitized' as much as possible to the variations in population density patterns. (This sensitization procedure can be seen as the method by which the gazetteer is transformed into the data-set used as a sampling-frame.)

By adjusting the gazetteer by the size of the at-risk population using census data in the manner outlined, there is an implicit caveat that population densities are uniform over EDs. This does not seem altogether likely, but it does seem reasonable to assume that rates will vary less across EDs individually, than they do over Tyne and Wear as a whole. Some improvement is therefore being made, although the boundary definitions of EDs, being somewhat arbitrary, may reduce the actual benefit, compared to its potential level. Alternatively, it can be argued that this arbitrariness implies a high level of self-cancelling in the error.

It may also be argued that the technique detailed above falls foul of the ecological fallacy (Openshaw 1985). This is the (fallacious) assumption that it is possible to infer down levels of aggregation from the spatial area (in this case the census EDs) to the individuals of that area (in this case the private households in that ED). Obviously, by admitting that the EDs involved are not internally homogeneous (all households do not have the same number of children resident, and so on) it becomes apparent that the procedure illustrated in Figure 8.2 may well give a grid reference as part of the control data-set that is not, in reality, home to any residents at all. This is not quite a fair criticism, however, because it slightly misses the point of the procedure. In actuality, Figure 8.2 does not attempt to ascribe the number of the population in a given household with absolute precision and accuracy. It does, rather, attempt to select households in a quasi-random fashion for later analysis. As such it is quite obviously not a perfect system, because it is working within the confines of a very limited data-set; the absolute accuracy of the technique in only selecting appropriately populated households is not as fundamentally important therefore, as if it were the actual body of the analysis itself. In short, while it is true that the 'sin' of the ecological fallacy is being committed, an alternative interpretation of the selection procedure is that it can be regarded as an imperfect method of choosing which households to use in comparison to the real data. The only alternative, that the existing densities are standard across Tyneside, which has already been rejected as being based upon fallacious assumptions, is less accurate and sensitive when compared to such data as exist.

The reliability of the approach just outlined is detailed in Raybould (1987).

THE SIGNIFICANCE TESTING PROCEDURE

Using the sampling-frame just created, cases were selected at random to reflect the number of actual cases in the data-set. This procedure was repeated 99 times, and the number of times the *control* cases' estimated mean pollution (see next section) level exceeded that of the *actual* cases was noted.

There was judged to be a significant difference between case and control cases if over 95 per cent of the control runs lay to one side of the actual case values for the test statistic. This is analogous to a probability of $p = 0.05$, in more conventional statistical techniques. In the case of this analysis, the test statistic was the (mean) average level of pollution displayed by the locations in question.

Before detailing the results of this test, it is first necessary to explain the nature of the other data which was used in this work, namely, the levels of heavy metals in the soils.

DATA ON THE HEAVY METAL VALUES

Introduction

The information on soil pollution levels was taken from an existing data-set which was augmented by extensive field-work in order to provide a more comprehensive coverage of the county. In the end, over 400 samples were taken at a basic density of one sample per square kilometre within Tyneside, with others located at random as a check upon the sampling and analysing procedures. The samples located specifically within grid squares (as opposed to being used to test the procedures) were located at random within it.

For each of these locations several heavy metals were tested for, and their levels recorded. These levels were then used in an interpolation procedure to create an estimated value at any given location such as those chosen as a control location in the Monte Carlo simulation, or noted as the actual location of a diabetic household. The variables analysed for this chapter were:

(i)	Cadmium	(total and available)
(ii)	Copper	(total)
(iii)	Manganese	(total)
(iv)	Nickel	(total
(v)	Zinc	(total)

This gave a total of five variables measured. The extra variable (available metal), which was assayed for cadmium, was a concession towards the argument that this chapter *should* make an attempt to evaluate the effects of metabolic pathways: whereas 'total' metal is exactly that, 'available metal' is the subset of that which is reasonably physiologically available for metabolic purposes. It is, essentially, that part of the 'total' measure which is less strongly chemically bonded to other parts of the soil which render the metal 'ineffective' as a potential chemical insult.

The interpolation of heavy metal data

Obviously, it is not possible to measure the amount of heavy metal in soil at *all* locations in Tyneside. It is therefore necessary to sample a set of locations within the county and use these to calculate what the probable levels of heavy metals are at those locations which were not measured directly. Such a procedure is known as 'surface interpolation'.

The LaPlacian Interpolation technique is an iterative procedure which ascribes to all points in its frame of reference the average of the adjacent points. This leaves an interpolated surface which is the smoothest which fits the data (Graham 1983) once the iterations have achieved convergence to a stable (and presumably unique) solution. In this way, the null hypothesis is the more likely outcome of a test, all other things being equal, as the resultant surface is flatter than might be expected. (This in turn implies that any positive results actually have a 'better' significance level than the literal ones quoted.)

Non-parametric correlation tests between the values of the metals at points on the interpolated surface generally show small but highly significant positive correlations. The one exception to this trend is total cadmium. It is not true, however, that there is therefore a general similarity of shapes to the surfaces. One possible explanation of this is that there is a degree of spatial variation, over the study area, of the level of association between the variables. The implications of making such an assumption are complex. For example, it may in turn imply the existence of spatial variations in the level of auto-correlation of one sample location to another across the study area. This would have implications in terms of the use of Kriging as an interpolation technique. (Kriging is an alternative interpolation methodology, based on the use of semi-variograms.) Alternatively, it may have a more direct form of influence upon the relationship between the patterns exhibited by the metals sampled and diabetes. What is likely, however, is that the amount of correlation between metals over the sample space will be reflected in the results of the investigation. If one metal is associated with diabetes it is commensurately more likely that the others will be so.

It should not be automatically inferred from this that the statistical validity of the results are questionable – at least on these grounds alone. Subsequent tests are independent of each other and the correlations are small (all less than 0.4). The possibility cannot be ignored, however.

While this may affect the ultimate possibility of claiming something to the effect that 'manganese *and* nickel *and* lead, *and* cadmium are *all* agents in the causation of diabetes', it does not have an effect upon the claim that there is a spatial association between heavy metals and diabetes in this context; heavy metal levels are regarded as a potential surrogate for general environmental conditions.

Results of the analysis

Whatever the implications of the correlations between the heavy metal variables, the results of comparisons between the real data and their Monte Carlo controls is shown in Table 8.3. (Results are shown beyond the level of resolution of the original assay techniques because the procedures in question all produce average calculations.)

The results show no particular associations between heavy metals and insulin-dependent diabetes. This implies that heavy metals are unlikely to have an immediate aetiological role, although further subanalysis would be required taking genetic facts into account to rule out a role in subpopulations. More important perhaps is the demonstration that this approach can be used to search for other potential environmental facts.

Ideally, such subsequent work should be based upon improved data: the use of interpolation procedures here was necessary to get around data restrictions, but obviously direct measurement is preferable. Monte Carlo is a good surrogate for a case control

Table 8.3 Results of the analysis

Metal	*Average for cases*	*Average for controls*	*Significance*	*Excess/ deficient*
Cadmium (available)	0.738306	0.7364307	–	–
Cadmium (total)	0.738706	0.738239	–	–
Manganese	299.21	308.11	0.10	Excess
Nickel (total)	5.37	6.02	–	Excess
Zinc (total)	143.98	144.56	–	Excess

approach, but it remains a surrogate nevertheless. The use of case controls would allow the residential migration history of children to be followed up to give a more accurate assessment of a child's history of exposure to heavy metals – or other environmental insults – and thereby avoid some of the more 'heroic' assumptions of this methodology.

In conclusion, this work has achieved the best that could be expected of it; its limitations lie largely in its very nature. The use of the Monte Carlo approach is required (given the nature of the data) and gives a rigorous explanatory overview of the situation. More sophisticated analysis awaits the application of alternative methodologies, which are not likely to be long in coming.

REFERENCES

Crow, Y.J., Alberti, K.G.M.M. and Parkin J.M. (1991) 'Incidence of childhood insulin-dependent diabetes mellitus in Northern England: relationship with social deprivation', *Brit. Med. J.* in press.

Graham, N. (1983) *A Combined Algorithm for Sample Design and Interpolation*, private publication.

Hope, A. (1968) 'A simplified Monte Carlo significance testing procedure, *J. Royal Soc. Stat.* Service B: 582–98.

Openshaw, S. (1985) *The Modifieable Areal Unit Problem*, Catmog 38, Norwich: Geo Books Ltd.

Raybould, S. (1987) 'The use of gazetteers in Geography: an example from epidemiology', *Northern Regional Research Laboratory*, Research Report no. 13, University of Newcastle.

WHO Study Group (1985) *Report on Diabetes mellitus*, WHO Technical Report Series 727, Geneva: WHO.

Chapter 9

Burning questions: incineration of wastes and implications for human health

Anthony C. Gatrell and Andrew A. Lovett

INTRODUCTION

The interest of geographers in the subject of hazardous waste disposal (Gatrell and Lovett 1986) has lagged behind their contribution to epidemiology, a discipline that seeks to describe and account for variations in ill health. Few attempts have been made to gauge the extent to which poorly-managed waste disposal may contribute to health problems. The present chapter discusses one method of waste disposal, incineration. It examines the types of incineration practised and what is currently known about the implications for human (and, briefly, animal) health. It then introduces a methodology that may be used to assess links between possible point sources of pollution, such as incinerators, and the incidence of a disease. This is applied to a case study that involves testing the hypothesis that cancer of the larynx is associated with proximity to the site of a former industrial waste incinerator. The need for better data on morbidity and mortality, and for high-quality environmental monitoring in the vicinity of some incinerators, is stressed.

Incineration is not the only form of waste disposal that may give rise to health problems. Landfill accounts for by far the greatest proportion of hazardous waste disposal, and if sites are poorly located with respect to underground hydrology, and leachate seeps into surrounding aquifers, then people may suffer from groundwater contamination. Given carefully-controlled site selection in Britain during the last fifteen years, this may be no cause for concern, but we raise the question as to whether older, unlicensed, and perhaps unrecorded, sites may create health problems in the future. There is limited evidence from elsewhere to suggest that contamination of groundwater by nitrates, due to poor waste disposal, may have led to raised incidence of neural tube defects (such as spina bifida and anencephalus) in the newborn (Dorsch *et al.* 1984). Hildyard (1983) has reviewed the problems of ill health thought to be associated with the Love Canal site near Niagara Falls (see Edelstein 1988, for an up-to-date discussion of the impact of this site). A further study in north-east USA (Budnick 1984) examined another landfill site but found no evidence of raised prevalence of congenital malformations. Anderson (1987) has a good discussion of the problems of evaluating such links.

INCINERATION

Incineration uses controlled combustion to dispose of a wide range of waste products, from garden wastes to human remains. A comprehensive examination of the process thus requires consideration, not only of major facilities for the burning of toxic wastes, but also municipal incinerators, hospital incinerators, and even crematoria. We direct our remarks in what follows primarily at sites licensed to dispose of toxic or hazardous wastes but will also comment on possible problems with hospital incinerators.

Land-based incineration of hazardous wastes in both Britain and the USA accounts for only 1–2 per cent of all such disposal (World Resources Institute 1987: 213; Hazardous Waste Inspectorate (HWI) 1985). In West Germany this proportion is about 15 per cent. An indication of absolute amounts incinerated (Table 9.1) makes it clear that France and West Germany burn between five and eight times as much toxic waste as does the United Kingdom, while the USA incinerates over thirty times as much.

Marine incineration of toxic wastes has been practised for several years. The UK incinerated on average 3,500 tonnes a year in the early 1980s, while Belgium and France burnt about 10,000 tonnes and the Netherlands about 20,000 tonnes per annum. West Germany disposed of circa 40,000 tonnes each year in the early 1980s (World Resources Institute 1987: 214). Recently, North Sea countries have agreed on a 65 per cent reduction in marine incineration by 1991 and a complete halt to such disposal by 1994.

Turning to commercial land-based incineration, there are currently four high-temperature plants in Britain. These are licensed to take in hazardous wastes from producers at home and abroad. Two are operated by the ReChem company and are located

Table 9.1 Average amounts of hazardous wastes incinerated on land in selected OECD countries, early 1980s

Country	*Amount (tonnes)*
Denmark	32,000
France	400,000
West Germany	675,000
Netherlands	66,000
Switzerland	120,000
United Kingdom	80,000
United States	2,700,000

Source: World Resources (1987).

at Pontypool, South Wales and Fawley in Hampshire. Until 1984 the company operated a third plant at Bonnybridge, near Falkirk in Scotland, but this was closed for economic reasons. A third currently operational plant is run by Cleanaway at Ellesmere Port in Cheshire. In 1987 a further plant, managed by Berridges at Hucknall in Nottinghamshire, closed after legal action by the County Council, instituted because the plant had not met planning requirements. The plant has been dismantled and relocated at Killamarsh in Derbyshire, where it is now run by Leigh Environmental.

The Hazardous Waste Inspectorate (before it became part of Her Majesty's Inspectorate of Pollution) viewed with concern in its Annual Reports the reduction in capacity of land-based incineration (Hazardous Waste Inspectorate 1986; 1988). The cessation of trading by Berridges in 1987 resulted in a temporary decline in capacity of about 10,000 tonnes and generated requests to HWI for further licences for marine incineration. However, during 1988–9 several companies announced plans to add to capacity. First, ReChem intend to expand operations at the Pontypool and Fawley plants (though these plans may have been affected by the decision of the Canadian government in late 1989 to end exports of polychlorinated biphenyls – PCBs – to Britain). Second, both Cleanaway and Ocean Environmental Management have announced plans to construct new incinerators at Ellesmere Port and Seal Sands (Teesside). Third, Leigh Environmental have made proposals for an incinerator near Doncaster. Last, International Technology Corporation have plans (jointly with Northumbrian Water) for incinerators on Tyneside and

Teesside). All these planned developments have run into opposition from either members of the local public, from local authorities, or from both (see, for instance, the *Guardian*, 27 April 1989).

What sorts of hazardous wastes are disposed of by the existing commercial incinerators? The ReChem plants incinerate high proportions of solid chemical wastes while the Ellesmere Port site burns only liquid wastes. The toxics tend to be organic chemical wastes produced in the manufacture of plastics and pharmaceuticals, but include pesticide residues and, notably, PCBs. The latter, generated as by-products in electrical industries, need to be incinerated at temperatures of circa 1,100–1,200 degrees Centigrade. Incomplete incineration can lead to the formation of dioxins and dibenzofurans, the toxicity of which are considered below. No evidence exists that the licensed merchant incinerators have failed to dispose properly of PCBs. Indeed, the Industrial Air Pollution Inspectorate conducted burning trials of PCBs at the ReChem and Cleanaway plants and demonstrated burning efficiencies of at least 99.999 per cent, with no detectable emissions of dioxins and dibenzofurans (Hazardous Waste Inspectorate 1985: 34).

In addition to these major sites there are about 100 smaller facilities for incinerating hazardous wastes that are generated 'in-house'. Such wastes include chlorinated solvents. Individual site licences give details of what such plants are authorized to dispose of and in what amounts, but the substances are described only in vague terms in a register of sites (Hazardous Waste Inspectorate 1984). The register also lists seventy-five sites licensed for the disposal by incineration of non-hazardous wastes. These include the municipal incinerators, managed by local authorities, which dispose of domestic wastes. Both the HWI (Hazardous Waste Inspectorate 1985: 34) and Hay (1982: 15–19) have suggested that such incinerators may emit dioxins and dibenzofurans.

Work in both the Netherlands and Switzerland confirmed that concentrations of both substances in the order of 0.1–0.2 ppm could be detected in the fly-ash of municipal incinerators, and concentrations three times as high as this in the fly-ash of a Swiss industrial heating plant. Further research, conducted in the Netherlands (Hay 1982: 18) found that when polyvinylchloride (PVC) and lignin were burnt together dioxins could be created. PVC is a widely-used product, employed in the manufacture of such things as gramophone records, plastic mats, wallpaper and adhesive tape. Given that lignin is present in wood and paper products it is not entirely surprising that dioxins might be found in municipal incinerator ash. These chemicals can be generated at relatively low temperatures (circa 600 degrees Centigrade) and if incineration is inefficient then toxic substances may be emitted, even though the materials feeding the incinerator are themselves harmless.

Some research has been conducted at the Warren Spring Laboratory on emissions of dioxins from municipal incinerators (Wallin and Clayton 1985). The authors looked at five incinerators and observed very high concentrations in one of these (Gateshead) but low concentrations elsewhere. They also modelled the dispersal of dioxins away from the stack of one incinerator (not Gateshead), and while there are no UK guidelines or air quality standards for dioxins, the maximum values observed were about 1/100th of limits for New York and about one-half of those for Ontario, Canada. It seems clear from other work (Clayton and Scott n.d.) that the Gateshead incinerator has not operated efficiently. Emission indices suggest that some dioxin compounds are present in concentrations up to 100 times greater than in another municipal incinerator studied by the authors. Ground-level concentrations predicted from a diffusion model are about 80–100 times higher. Clayton and Scott call for further work to identify the reasons for the high dioxin (and also particulate) emissions from the Gateshead plant.

Her Majesty's Inspectorate of Pollution has undertaken a study of background levels of various dioxins and dibenzofurans in the environment. Their results, comprising samples taken at seventy-seven sites forming the intersections of a regular 50 km grid across England, Wales, and parts of Scotland, suggest that the various compounds are ubiquitous (Hazardous Waste Inspectorate, 1988: 55). The published figure will enable some comparisons to be drawn with results from soil surveys, currently underway, in the vicinity of various types of incinerators.

The last HWI report was highly critical of hospital incinerators, both with respect to the storage of wastes and the operation of the plants. 'If the, admittedly small, number (15) of hospitals visited are typical of the national picture then the situation is deplorable' (Hazardous Waste Inspectorate 1988: 14). The National Society for Clean Air (NSCA) (1988) has conducted its own investigation of such facilities, including not only hospital incinerators, but also other local authority facilities protected from prosecution because of Crown immunity. The NSCA survey of 483 Environmental Health Officers (EHOs) generated a response rate of 68 per cent.

About one-quarter of the 1,700 boilers and incinerators protected by Crown immunity were considered by EHOs to be sources of air pollution. Thirty-nine EHOs expressed concern about levels of dioxins and dibenzofurans, partly because of the high plastic content of hospital waste, though none had any detailed figures to support these concerns. The NSCA compiled a 'dirty dozen' of local authorities that included the worst offending hospitals. These areas were: Bradford, Camden, Canterbury, Cardiff, Dundee, Gateshead, Greenwich, Hemel Hempstead, Kettering, Norwich, Sheffield and Welwyn.

INCINERATION AND HEALTH

Given that incineration may be a source of environmental pollution, what evidence exists that such pollution may be detrimental to health? In particular, is there any evidence to support the view that dioxins and related compounds are carcinogenic or teratogenic (causing congenital malformations)?

Experiments with rodents suggest that dioxins are potent teratogens (Fletcher 1985: 86), but the evidence concerning human health is less clear. The explosion at Seveso, Italy, in 1976, discharged about 250g of dioxins into the atmosphere and led to severe skin disfigurement (chloracne) in a number of local residents. Evidence on prevalence of congenital malformations is inconclusive and bedevilled by the usual problems of variable reporting. For instance, the malformation rate in the Seveso region rose from 1.03 per 1,000 births in 1976, to 13.7 in 1977 and 19.0 in 1978, but Hay (1982: 216) suggests that underreporting before the disaster was widespread. A cancer registry has been established in the region, but Hay (1982) argues that any impact on cancer rates will not be revealed for many years because of the long latency period between exposure and tumour development and diagnosis. It is thus of particular interest to note a report of research in West Germany conducted into the incidence of cancer among workers exposed to dioxins in an industrial explosion in Ludwigshafen in 1953 (Yanchinski 1989). Observed mortality due to cancer of the larynx, trachea, bronchus and lung among these workers was statistically in excess of expected numbers. This study appears to be the first to document a link between dioxin exposure and cancer.

Gough (1986) has examined dioxin contamination due to the use of 'agent orange' as a defoliant in the Vietnam War. Worries have been expressed about the prevalence of congenital malformations in offspring of both the indigeneous population and American servicemen exposed to the chemicals. In a careful case-control study Erickson (1984) showed that Vietnam veterans who were probably exposed to agent orange had a greater risk of fathering babies with spina bifida. There was, however, no raised risk of the related defect anencephalus. Gough takes a sceptical view of the teratogenicity of dioxins. A well-known environmentalist, in a review of the book, concludes that 'concern about the release of dioxin during the incineration of plastics in municipal incinerators is probably misplaced' (Pearce 1986: 50).

The most widely-publicized of the possible links between incineration and ill health is that relating to the ReChem plants at Bonnybridge and Pontypool. Cases have been reported in national newspapers of congenital eye malformations in children born in the vicinity of the two plants. These defects include the absence of an eye (anophthalmos) and a reduction in the size of an eye (microphthalmus). Government studies have been conducted by both the Welsh and Scottish Offices and neither found any evidence to support the assertions of a raised incidence of eye malformations in the vicinity of the ReChem plants. Indeed, the Welsh Office report (Welsh Office 1985) found, over a ten-year period (1974–83), no cases of eye malformations in Torfaen district, within which the Pontypool plant is located. This appeared to be at variance with some cases identified and documented in the public press.

Part of the explanation may lie with the main source of government statistics, assembled by the Office of Population Censuses and Surveys (OPCS) from notifications made by district medical officers. These notifications are voluntary and it is quite possible for genuine cases not to be registered with OPCS. In some parts of the country (Birmingham, for example) there are local registers of congenital malformations and comparisons of these data with those from OPCS show the latter to be of dubious quality when some classes of malformation are considered. Knox and his co-workers have gone so far as to argue that OPCS data on 'the malformations of the eye ... must be regarded as almost entirely useless, by reasons of late diagnosis, inaccurate reporting and their generally trivial nature' (Knox *et al.* 1984: 303).

The Hazardous Waste Inspectorate found the official reports reassuring, suggesting that they 'reinforce repeated statements by the regulatory authorities and plant operators that these facilities are operating safely and efficiently' (Hazardous Waste Inspectorate 1985: 34). Such reassurance does not appear to be shared by some local residents. Phipps (1986) conducted a detailed survey of attitudes to the Pontypool plant and found that about three-quarters of a sample of nearly 800 residents expressed opposition to it.

Concerns about the impact of hazardous waste incineration on animal health have also been voiced, and addressed in part by the report on the now-closed Bonnybridge plant (Lenihan 1985). Farmers in the Bonnybridge area had reported raised levels of abnormalities, stillbirths and unexpected deaths in cattle, but the Lenihan report was unable to confirm any link between environmental pollution and health problems with cattle. However, researchers at Dundee University have been critical of the official report and have argued that the combined effects of heavy metal pollution and polychlorinated hydrocarbons were not properly assessed (Lloyd *et al.* 1988). Their own work has indicated very high levels of twin births, in both cattle and human populations, in the areas most at risk from air pollution. More detailed studies of soil contamination (especially by chromium) have shown raised levels of heavy metal pollution in a field, close to the site, grazed by cattle (Smith and Lloyd 1986). Researchers from the company have suggested that deaths of cattle were due instead to poisoning by the ragwort weed (Eduljee 1986), though this has been contested (Lloyd *et al.* 1987).

A final piece of evidence that suggests there may be a link between incineration and ill health comes from work in Newcastle on childhood leukaemia (Openshaw *et al.* 1987). This research, as well as confirming the existence of a 'cluster' of cases near the Sellafield nuclear reprocessing plant, has indicated the presence of a further cluster in Gateshead, a finding which has been linked tentatively to the presence there of a municipal incinerator. It should be recognized that the methodology adopted by Openshaw and his colleagues has not gone uncriticized. Indeed, it is welcoming to see new methodologies being developed by statisticians to detect and model spatial patterns of rare diseases (Hills and Alexander 1989), one approach to which we now outline and illustrate.

LARYNGEAL CANCER IN A LANCASHIRE HEALTH AUTHORITY

We turn now to an empirical study that seeks to assess whether there is any association between the geographical distribution of a specific cancer and proximity to an industrial incinerator. We first describe the background to this study, then outline in brief form our method for tackling the research and the findings that have emerged.

An industrial waste incinerator operated near the small town of Coppull in Lancashire between 1972 and 1980. It was used primarily for the disposal of liquid wastes, mostly solvents and oils. During the lifetime of the plant there were frequent public protests about irritant gases produced during combustion; since its closure there have been calls for research to monitor longer-term health effects.

Data were made available by the Health Authority on cancers registered between 1974 and 1983 and we undertook some simple descriptive mapping to see if there was any visual evidence of localized clustering of cases in the vicinity of the plant. Such mapping

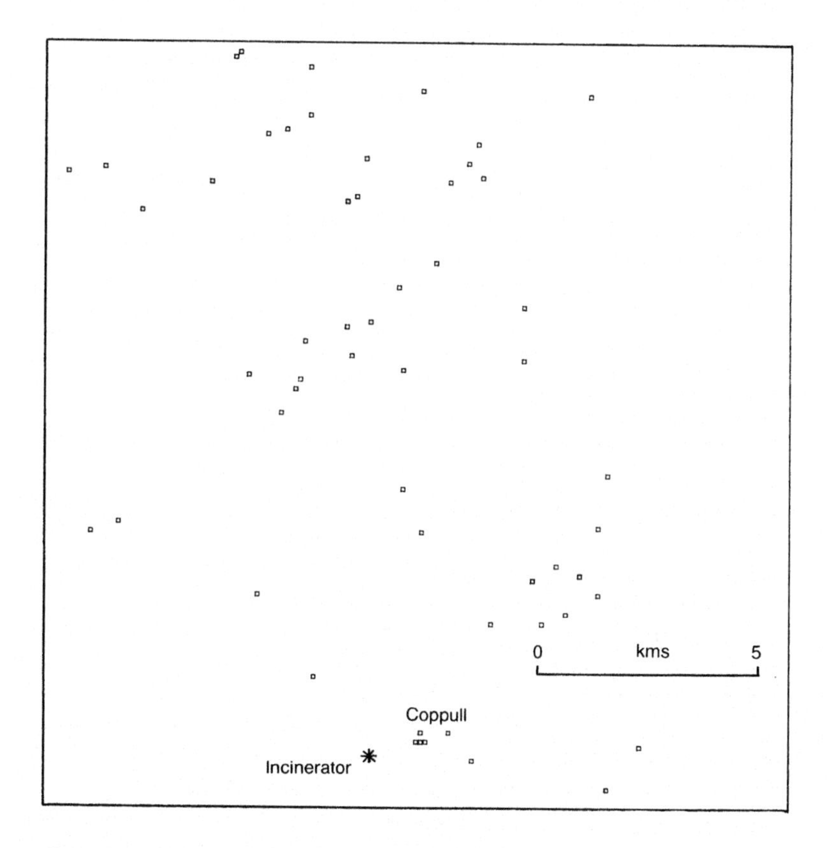

Figure 9.1 Distribution of cancer of the larynx

was made possible by converting the unit postcodes of cases to Ordnance Survey grid references, accurate to 100 metres; the computerized Central Postcode Directory was used for this purpose. From a purely subjective point of view it appeared that the distribution of most cancers mirrored the distribution of population as a whole. The exception to this was laryngeal cancer. Here, out of only fifty-eight cases notified during the study period, five were observed in Coppull, four of whom lived in the south-west of the town within two kilometres of the closed plant (see Figure 9.1).

We reported on this finding at the Institute of British Geographers' Annual Conference in January 1989, suggesting that the distribution was sufficiently unusual as to warrant analysis rather than subjective assessment. Reports on the conference presentation in the national press and on radio highlighted our preliminary finding and led to a spate of often alarming local reports on the possible problem. More positively, contacts were established with community physicians and Environmental Health Officers elsewhere in the country, who were themselves concerned about possible risks from proximity to incinerators.

We therefore resolved to perform some rigorous statistical analysis of the data and this has now been accomplished. An outline of the method we used is described in Diggle *et al.* (1990), although the full statistical argument is available elsewhere (Diggle 1990). Briefly, it entailed setting up a formal statistical model which stated that the 'intensity' of cases of laryngeal cancer is a function first of 'background' intensity (population at risk) and, second, of distance from the putative source of pollution. The model tested whether proximity to the pollution source is a significant influence on the distribution of laryngeal cancer.

The method therefore required information on 'background intensity'. Given that we had point data on the distribution of cases (Figure 9.1) we wished to avoid using data for what are essentially arbitrary spatial units such as electoral wards or Enumeration Districts. Ideally, we required information that reflected population distribution but which was not associated with the putative source. In the absence of point data on population distribution we initially used data on the distribution of the much more common lung cancer, of which there were 978 cases over the study period. This had the advantage that a known risk factor, tobacco smoking, is common to both cancers (Alderson 1986). It also had the advantage that the spatial distribution (see Figure 9.2) seemed to reflect quite well the natural tendency of population to cluster into towns and villages. A further advantage was that the age-sex structures of the two diseases are very broadly, though not in detail, comparable. The principal disadvantage was that there was no guarantee that lung cancer was not itself associated with proximity to the incinerator. We overcame this problem by experimenting with other measures of background intensity.

The statistical model was fitted using the method of maximum likelihood. A 'null' model was fitted, which assumed that proximity to the incinerator had no effect and that cancer of the larynx follows the same spatial distribution as lung cancer. This was then compared with a maximum likelihood estimate that incorporated the parameters of a distance function. In essence, the difference

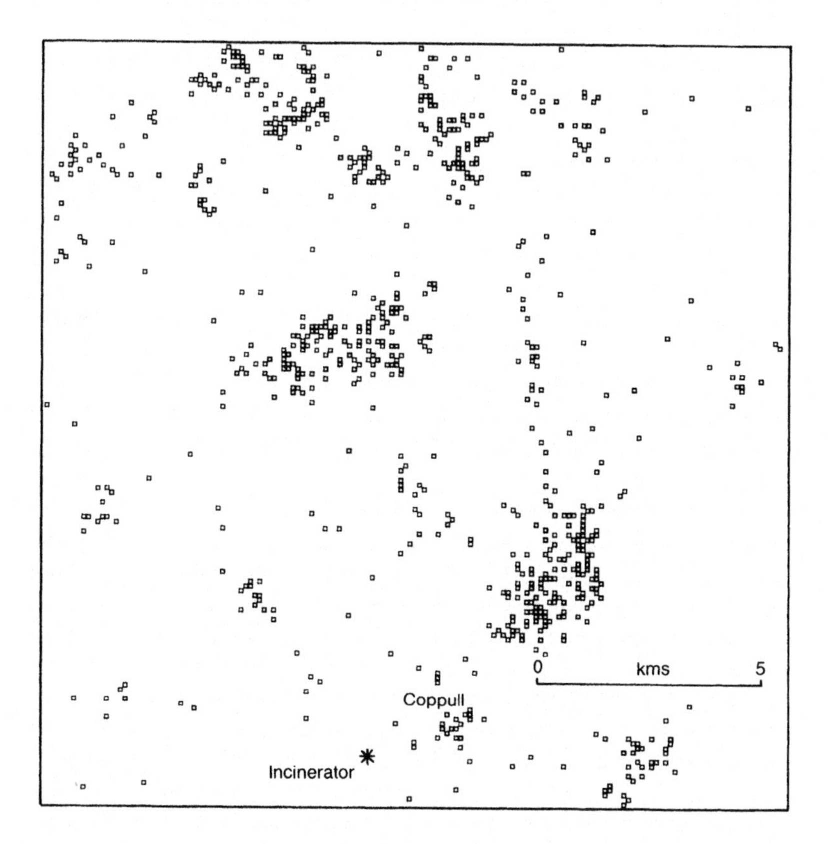

Figure 9.2 Distribution of cancer of the lung

between the two likelihood values (with and without a distance effect) provided a test of the null hypothesis that distance has no effect.

We found that there is a statistically significant association with proximity to the incinerator. Repeating the analysis with another cancer, stomach, for which there were 398 cases, led to the same conclusion. We have also sought to use a measure of background intensity that is unrelated to cancer incidence. This has involved taking all unit postcodes within the District Health Authority and using these as a 'control' surface. There are 4,389 such postcodes, each representing on average sixteen households, though there is likely to be considerable variation about this mean. The distribution of such postcodes provides a reasonable map of population distribution. Using this as background intensity again confirms the earlier finding.

Lastly, we have posed the following question. Is the location of the closed incinerator the only site within the district that has an 'influence' on the spatial distribution of laryngeal cancer? We have answered this question by 'moving' the location of the putative source across the map on a regular grid and fitting the model each time to the 'new' source. Regardless of where we move the incinerator, the only location to which laryngeal cancer intensity is linked is the actual site.

Further details of the results described above are given in Diggle *et al.* (1990) and in Gatrell (1990). In particular, most recent work has widened the study area, since it could be countered that the results depend upon the location of the incinerator in the south-west of the district. Initial results (Gatrell 1990) suggest that earlier conclusions remain valid.

Clearly, in the absence of additional information it is foolish to claim that living near the incinerator has 'caused' cancer of the larynx. We do not have information on the residential histories of the cases, or details of where they worked. (Interestingly, however, because of the publicity generated by our preliminary work, anecdotal information was received from individuals whose relatives had formerly lived in Coppull and who had now contracted the disease). We have no information on the latent period, inevitably variable for different individuals, over which the cancer will have developed. This may have been five years (which would of course strengthen the argument for causation) or twenty-five years (well before the incinerator commenced operation). Further, we have no details about smoking histories, alcohol consumption, or other risk factors for individuals. Nor do we have background environmental information on air pollution which should have been monitored in detail at the time of operation. This said, we do believe that the methodology is a valuable one and could be used in other investigations into possible links between ill health and incineration.

CONCLUSIONS

If we are to make progress in assessing possible links between incineration and human health we need access to better data. If there is indeed a link with cancer of the larynx then such a link may take years to manifest itself. It may be more profitable to look for possible links to respiratory diseases and to monitor attendances at general practitioner surgeries. Even here, of course, the assumption would be made that complaints about illness brought to the GP were an adequate reflection of morbidity in the population at large.

We have also highlighted the need for better data on congenital malformations if there is to be any progress in studying possible environmental factors in the aetiology of eye defects. To this end, we are engaged in setting up the first national prospective study of eye malformations in children. This is being done in collaboration with consultant surgeons and ophthalmologists at Moorfields Eye Hospital in London. Given the rarity of the conditions it will take time to collect a reasonable body of material, but perhaps within five years we shall be in a position to paint an adequate picture of such morbidity and to test hypotheses about causation. We cannot fail to reach the conclusion that some previous work has relied on anecdote and speculation and that we need to bring to bear to such problems the combined skills of epidemiologists, geographers and statisticians in researching this important area of public concern.

REFERENCES

Alderson, M. (1986) *Occupational Cancer*, London: Butterworths.
Anderson, R.F. (1987) 'Solid waste and public health', in M.R. Greenberg (ed.) *Public Health and the Environment: the U.S. Experience*, pp. 173–204, London: Guilford Press.
Budnick, L.D. (1984) 'Cancer and birth defects near the Drake Superfund site, Pennsylvania', *Archives of Environmental Health* 39, 409–13.
Clayton, P. and Scott, D.W. (n.d.) *The Measurement of Suspended Particle, Heavy*

Metal and Selected Organic Emissions at Gateshead Municipcal Refuse Incinerator, Hertfordshire: Warren Spring Laboratory.

Diggle, P.J. (1990) 'A point process modelling approach to raised incidence of a rare phenomenon in the vicinity of a pre-specified point', *Journal of the Royal Statistical Society* Series A, 153, 349–62.

Diggle, P.J., Gatrell A.C. and Lovett, A.A. (1990) 'Modelling the incidence of cancer of the larynx in part of Lancashire: a new methodology for spatial epidemiology', in R.W. Thomas (ed.) *Spatial Epidemiology*, London: Pion, 33–47.

Dorsch, M.M., Scragg, R.K., McMichael, A.J., Baghurst, P.A. and Dyer, K.F. (1984) 'Congenital malformations and material drinking water supply in rural South Australia: A case-control study', *American Journal of Epidemiology* 119, 473–86.

Edelstein, M.R. (1988) *Contaminated Communities: the Social and Psychological Impacts of Residential Toxic Exposure*, Boulder, Col: Westview Press.

Eduljee, G.H. (1986) 'Soil pollution at Bonnybridge', *Chemistry in Britain* April, 308–9.

Erickson, J.D. (1984) 'Vietnam veterans' risks for fathering babies with birth defects', *Journal of the American Medical Association* 252, 903–12.

Fletcher, A.C. (1985) *Reproductive Hazards of Work*, London: Association of Scientific, Technical and Managerial Staffs, Equal Opportunities Commission.

Gatrell, A.C. (1990) 'On modelling spatial point patterns in epidemiology: cancer of the larynx in Lancashire, North West Regional Research Laboratory, Research Report No. 9.

Gatrell, A.C. and Lovett, A.A. (1986) 'Hazardous waste disposal in England and Wales', *Area* 18, 275–83.

Gough, M. (1986) *Dioxin, Agent Orange: The Facts*, New York: Plenum Press.

Hay, A. (1982) *The Chemical Scythe: Lessons of 2,4,5,-T and Dioxins*, New York: Plenum Press.

Hazardous Waste Inspectorate (1984) *Register of Facilities for the Disposal of Controlled Wastes in England and Wales*, London: Department of the Environment.

Hazardous Waste Inspectorate (1985) *Hazardous Waste Management: An Overview*, First Report of the HWI, London: HMSO.

Hazardous Waste Inspectorate (1986) *Hazardous Waste Management: 'Ramshackle & Antediluvian?'*, Second Report of the HWI, London: HMSO.

Hazardous Waste Inspectorate (1988) *The Hazardous Waste Inspectorate*, Third Report of the HWI, London: HMSO.

Hildyard, N. (1983) *Cover-up*, Sevenoaks, Kent: New English Library.

Hills, M. and Alexander, F. (1989) 'Statistical methods used in assessing the risk of disease near a source of possible environmental pollution: a review', *Journal of the Royal Statistical Society*, Series A, 152 (3), 353–63.

Knox, E.G., Armstrong, E.H. and Lancashire, R. (1984) The quality of notification of congenital malformations, *Journal of Epidemiology and Community Health*, 38, 296–305.

Lenihan, J. (1985) *Bonnybridge/Denny Morbidity Review*, Edinburgh: Scottish Home and Health Department.

Lloyd, O.L., Smith, G.H., Lloyd, M.M., Williams, F.L.R. and Hopwood, D. (1987) 'Bonnybridge revisited', *Chemistry in Britain* January, 31–2.

Lloyd, O.L., Lloyd, M.M. Williams, F.L.R. and Lawson, A. (1988) 'Twinning in human populations and in cattle exposed to air pollution from incinerators', *British Journal of Industrial Medicine* 45, 556–60.

National Society for Clean Air (1988) *Air Pollution from Crown Property*, Brighton: National Society for Clean Air.

Openshaw, S., Charlton, M., Wymer, C. and Craft, A. (1987) 'A Mark I Geographical Analysis Machine for the automated analysis of point data sets', *International Journal of Geographical Information Systems* 1 (4), 335–58.

Pearce, F. (1986) 'Review of M. Gough: dioxin, agent orange: the facts', *New Scientist*, 5 June, 50.

Phipps, S. (1986) The public and the polluter: a local perspective on Re-chem's plant at Pontypool', Working Paper no. 467, School of Geography, University of Leeds.

Smith, G.H. and Lloyd, O.L. (1986) 'Soil pollution from a chemical waste dump', *Chemistry in Britain* February, 139–41.

Wallin, S. and Clayton, P. (1985) *Emissions and Pathways of Dioxins and Dibenzofurans from the Combustion of Municipal Refuse and Refuse-derived Fuel*, Hertfordshire: Warren Spring Laboratory.

Welsh Office (1985) *The Incidence of Congenital Malformations in Wales, with special reference to the District of Torfaen, Gwent*, Cardiff: Welsh Office.

World Resources Institute (1987) 'Managing hazardous wastes: the unmet challenge', *World Resources* 201–19.

Yanchinski, S. (1989) 'New analysis links dioxin to cancer', *New Scientist* 28 October.

Chapter 10

Assessing the health effects of waste disposal sites: issues in risk analysis and some Bayesian conclusions

Trevor A. Sheldon and Denis Smith

INTRODUCTION

> One of the most attractive features of the Bayesian approach is its recognition of the legitimacy of plurality of (coherently constrained) responses to data. Any approach to scientific inference which seeks to legitimise an answer in response to complex uncertainty is, for me, a totalitarian parody of a would-be rational human learning process.
>
> (Smith 1984)

Increasingly the public is becoming aware of environmental issues, such as the potential hazards to health resulting from pollution and from the disposal of a range of wastes. In particular, there is great anxiety about the environmental risks associated with nuclear power generation, the reprocessing of nuclear waste and the disposal of both nuclear and non-nuclear waste products. 'Green' issues have been pushed to the centre of the political stage and intense debate rages about the existence, location and regulation of the disposal of a range of waste streams. Much of this discourse focuses on the question of the identification and quantification of the health risks associated with such activities. In particular, uncertainty exists over whether there are raised health risks associated with various forms of waste disposal, their nature and extent, how local policies are related to exposure and, lastly, the contentious issue of risk management. The evaluation of the various disposal options such as landfill, incineration and ocean dumping have to take into account the damage they may cause to the health of present and future generations.

This paper seeks to address these issues and explores their

relevance for a geographical appreciation of the waste problem. In addition, a Bayesian perspective is introduced drawing on examples from disease mapping and epidemiological studies which examine the potential health effects of waste disposal sites. The argument is then developed that when epidemiological investigations are made in areas with an obvious policy implication then traditional statistical perspectives are not very helpful in moderating between competing models, results or opinions. Nor do the traditional models fit well into the decision-making process where one has to consider the problems of relating science to policy. Furthermore, it is considered that a Bayesian approach to scientific reasoning provides a more natural framework for both biostatisticians and social scientists when considering risk assessment and management. The proposition then is that geographers who are working on environmental hazards to health (as, indeed, those working in other areas) should seriously consider adopting a Bayesian perspective which may prove useful in unifying some of the methodological and normative aspects of their work.

A common procedure in researching links between potential risk factors and disease outcomes is the simultaneous examination of large numbers of associations. Since there are underlying geographical variations in risk factors arising from industrial activity this may be expected to manifest itself in spatial variation in the incidence of disease. This is not a new discovery and there is a rich tradition of disease mapping in epidemiology. For example, John Snow used dot maps to reveal the topography of the cholera outbreak in the Soho area of London in 1854 and to draw inferences about the source of the infection (Cliff and Haggett 1988). Epidemiology is unique in its attempt to bring together and analyse as a whole, the multiplicity of potential explanatory factors which converge to cause disease outbreaks. Environmental epidemiology is itself an inherently interdisciplinary pursuit and has contributed to the identification and subsequent control of hazards such as asbestos (Doll 1955).

Put simply then, the question that we are trying to answer within the framework of this chapter is as follows: what are the health risks of proximity to a waste disposal site and how should that information be used in the decision processes concerning the possible closure of a facility or in the formulation of general policy on the location and management of waste disposal sites. This chapter attempts to review some of the problems in the assessment of health risks associated with the disposal of waste, and questions the more traditional statistical approach used by epidemiologists and geographers alike.

There are a number of major issues facing geographers when they embark upon this type of research. The first relates to methodological questions concerning epidemiological approaches to environmental risk analysis such as exposure assessment, cluster techniques, and so on. The second issue concerns those problems associated with multiple inference and mapping, especially when dealing with rare diseases. Finally, problems inevitably exist in using research results which are typically found at the interface between science and policy. This generates conflict between expert groups and clouds the issue of public participation in decision-making. Geographers, as social scientists, have the potential to play a role in appreciating and developing the science-policy interface, more fully perhaps, than practitioners with a biostatistical background who may have no training in policy analysis. The remainder of this chapter seeks to examine these issues further, starting with an analysis of the methodological issues involved in environmental risk assessment.

METHODOLOGICAL PROBLEMS IN ENVIRONMENTAL RISK ASSESSMENT

There are a number of problems inherent in the use of risk assessment within public policy-making, not the least of which is the degree of uncertainty that inevitably surrounds such analyses. Considerable attention has been focused on the role of science within decision-making for hazardous activities and the main conclusion from such work has been that science can heighten conflict whilst seeking to ameliorate it (see, for example, Nelkin 1975; Wynne 1982; Irwin 1985; Collingridge and Reeve 1986; Smith 1990). The essence of the problem lies in the reluctance of experts to acknowledge the bounds of their own knowledge and this exposes them to criticism from other experts. The resultant disagreement causes public confidence in expert prognosis to be severely undermined and often results in a heightening of concern. Within the area of waste disposal the problem is made more acute by the dearth of available data in the public domain concerning the potential nature and range of environmental impacts which result from waste treatment and disposal.

There is, therefore, an obvious need to both challenge basic assumptions about the nature of waste-related hazards and to define initial assessments in the light of new data. The latter point is illustrated by the recent recognition that landfill, once considered a 'final solution' to the waste issue, is little more than a holding option as wastes re-emerge at a later date into the ecosystem. Indeed, landfill sites, along with incinerators, have provided much of the early empirical data in the field of health effects (see Stern *et al.* 1989; Griffith *et al.* 1989b; Heath 1983; Levine and Chitwood 1985; Thomas 1988) with Love Canal providing a 'convenient' source of the initial data (see Vianna and Polan 1984).

Despite this work there are still a number of methodological issues which need to be addressed. With these in mind, Anderson (1985) has provided a set of criteria for successful waste site epidemiological studies. We can now consider some of these issues prior to examining the potential benefits associated with a Bayesian approach to risk assessment.

Exposure assessment

The corner-stone of any epidemiological study is the assessment of exposure. However, in environmental epidemiology this poses one of the most difficult questions. Most epidemiology relies on an assessment of 'naturally' occurring exposures which result from occupation, residence and the like, but accurate exposure data rarely exist. In addition, those exposures which can be observed are rarely random: occupations, residence and human behaviours are imposed or self-selected in a non-random way such that it can be difficult to tease out the independent effects of exposure, since other attributes differ between exposed and non-exposed populations.

There is a need, therefore, for more systematic assessment, collection and recording of information about human exposure to environmental risk factors. This needs to be done in the context of developing data bases and record linkage systems so that more comprehensive information can be obtained. The potential for Geographical Information Systems (GIS) here has yet to be developed, although it is apparent from a number of early studies that the techniques could play a useful role within epidemiology.

The difficulty and, at times impossibility, of accurately estimating direct exposure levels to potentially hazardous chemicals from waste sites has made it difficult to carry out good quality and influential epidemiological work in this area (Heath 1983). However, recent developments in the use of biochemical and biological markers to estimate the biologically effective dose, offer considerable potential to improve such analyses (Griffith *et al.* 1989a). For example, polychlorinated biphenyls (PCBs) can now be monitored by testing for PCB residues in human tissues and breast milk. This has obvious implications for assessing the health impacts arising from industrial high-temperature chemical incineration.

The use of biological markers in risk assessment may enable better modelling of dose-response relationships, improve the evaluation of past doses in epidemiological studies and perhaps help in the identification of previously unrecognized genetic hazards (Hattis 1988). However, the technique is of no help in those situations where there are several point sources of pollution in a relatively small area and one is trying to assess the risks associated with them separately.

Data sources, hypothesis generation and outcome definition

Where epidemiology does have significant strength is in testing a specific hypothesis, such as whether a particular disease is related to a given exposure. However, it becomes increasingly difficult to answer the more general question of whether there are any adverse health effects associated with a given exposure.

At the present moment there is a very poor degree of environmental disease surveillance carried out in Britain. There is, for example, no systematic regional or national framework for the reporting of environmental illness. As a consequence, Anderson (1985) proposed that a list of 'sentinel' diseases be developed which can link up with small area geographical co-ordinate systems. Such an approach has a number of obvious merits in the provision of quality epidemiological data. However, it requires more 'bespoke' studies and detailed surveillance, which would need to be carefully designed in order to investigate suspected sites of exposure and likely toxicants. The problem with many previous studies is that they attempted an analysis a few years after exposure to the hazard ended (Levine and Chitwood 1985). In addition, they have usually been based on a small population sample which resulted in too few cases to allow for the detection (as statistically significant) of potentially important health effects.

There is a requirement, therefore, for a systematic collection of data which would include birth and disease registers and the monitoring of levels of pollutants both at work and in areas near suspected point sources of pollution. The development of cancer registries producing diagnostically reliable and complete data sets will help further aetiological research. It is to be hoped that other morbidity registers, perhaps generated through primary care facilities, will develop further and thereby contribute to the much-needed data base. Registers linked through primary care facilities could also be used to produce community morbidity profiles, along the lines of those in Cuba, thereby 'providing a data base which resembles an epidemiologist's dream' (Stubbs 1989: 94–5). The role of GIS within this context is extremely important as it could provide the means for handling such large data sets and giving them a spatial structure.

The coverage of the data is obviously an important aspect of such work but, for geographers, one of the vital demands is for the data to be spatially referenced. The lack of spatial detail and consistency of official sources of health data currently impedes the use of powerful Geographical Information Systems for health research (Twigg 1990; Carstairs and Lowe 1986). If good data of this sort existed, it could be linked up with other digitally-stored geographical information on topology, drainage patterns, soil distributions, transportation routes and other relevant environmental information, which could then be used for the purposes of monitoring and co-ordinating the activities of hazardous waste sites (Estes *et al.* 1987).

However, at the moment the available data are generally of a poor quality. Aggregate data sources are likely to be incomplete, will usually contain errors and misclassifications and will inevitably be out of date. The variable quality of such data limits our ability to make comparisons between communities, both spatially and temporally, thereby introducing a high degree of uncertainty into the analysis. Such uncertainty is not so common in experimental toxicology research which undertakes quantitative risk assessment on animals (Stallones 1988), although in this case uncertainty exists as we try to extrapolate such findings to humans.

Cluster analysis

One of the most significant problems facing waste-related health studies can be found in terms of the spatial limits that need to be imposed on the analysis. The hazard range of any given pollutant will be a function of the mode of release, the population at risk (determined as a function of population density), the anticipated exposure routes (for example, ingestion, dermal absorption, particulate inhalation), uptake in water courses and the food chain and atmospheric factors (wind speed and direction, pollutant plume behaviour, rainfall, and so on). If we examine a range of areas for a variety of potential health effects then this creates a problem of multiple significance testing which is common in this type of work. Finally, the issue of causality is further complicated when assessing data from an area with multiple sources of pollution. Since individual exposures are so difficult to ascertain, the spatial distribution of disease around a potential source of pollution is often used as a proxy. If exposure is associated with risk then one would expect the incidence to be raised in populations close to the waste site. The detection of clusters of disease has thus been a major preoccupation for both epidemiology and medical geography.

In the past, the identification and investigation of clusters has contributed greatly to our knowledge of health risks (Rothenberg *et al.* 1990). For example, within the identification of developmental risks, the teratogenic effects of methylmercury, PCBs, thalidomide and diethylstilbestrol were all identified in this way (Hogue and Brewster 1988). However, many reported clusters turn out, on further investigation, to be explainable by the distribution of other known risk factors such as age, class and parity. Thomas (1985), for example, cites a case of a false positive cluster which was presented as a reported epidemic of cancer in a particular street in the Chomedy district of Montreal during 1981. The outbreak suggested an unknown environmental cause and was subjected to a rigorous epidemiological study. This analysis involved careful definition of the population and period of time at risk, the selection of suitable control populations for comparison, a complete ascertainment of the cases and confirmation of diagnoses, and conclusions were that no excess of cancers had indeed occurred (Spitzer *et al.* 1982). Cases such as this raise questions about the use of 'anecdotal evidence' in identifying a cluster.

On the other hand, clusters of disease in time or space may be difficult to identify due to the background levels of exposure in individual cases. It is easier to identify hazards by cluster analysis when there are either epidemics, such as the thalidomide babies or the methylmercury disaster in Minamata Japan, or very unusual illnesses, such as the cases of vaginal adenocarcinoma among young women associated with maternal consumption of diethylstilbestrol. It is much more difficult, however, to identify an increase in the incidence of low birth-weight babies following exposure to waste streams, for example, which is a much more common phenomena. This difficulty is illustrated by the example of Love Canal in New York State, where there was an apparent rise in low birth-weight among resident births around the chemical dump site during a period of active waste dumping by the Hooker Chemical Company in the 1940s (Vianna and Polan 1984; Goldman *et al.* 1985). If this is a real effect, then again causality is difficult to prove as studies were only initiated in the late 1970s when foul-smelling liquids and sludge seeped into the basements of houses built on top of the dump. It should be noted that only local residents appear to have been at risk, and so surveillance based on community or county rates would not have revealed the 'epidemic'. Thus it is only by some *a priori* assessment of the geographical areas likely to be at risk that one is likely to be able to initiate meaningful studies during the pollution episode (see, for example, Hogue and Brewster 1988).

There are, therefore, fundamental difficulties in carrying out and interpreting such aggregate data in ecological studies. Even after standardizing for known demographic variations of age and sex between regions, all sorts of other known and unknown confounders can operate to distort the analysis. Exposure may vary greatly within the geographical areas examined and so might dilute the findings. In addition, there may be regional differences in the reporting and classification of disease and the denominator of the rates, that is, the population at risk, may be incorrect (Lilienfeld and Lilienfeld 1980). For these reasons many researchers attempt to focus on individuals rather than aggregates, either by means of a cohort with suitable control areas, or a case-control study (Gardner *et al.* 1990; Lovett *et al.* 1990). However, it is still important that there is an appropriate selection of controls and that, in the case of aetiological studies of clusters, close matching for location is carried out (Breslow and Day 1980).

No discussion of disease cluster detection in a geography text would be complete without mention of the work of Openshaw and colleagues at Newcastle University. In their work to identify leukaemia clusters, Openshaw *et al.* (1987, 1988), used the geographical analysis machine (GAM) which generates multiple overlapping circles of varying sizes. Using estimates of the population contained in each circle from small area statistics, the actual number of leukaemia cases (points) inside the circle is compared to that which might be expected given a null hypothesis of a random

(Poisson) distribution of cases. Those circles which have significantly more leukaemia cases than expected are drawn on a map. This rather mechanical and simplistic approach has been justifiably criticized elsewhere (Clayton 1988; Hills and Alexander 1989) and also suffers from problems which are commonly found in other epidemiological studies. These include multiple significance testing, poor definition of what constitutes a cluster, confusing statistical significance with public health significance and poor adjustment for confounders.

With these points in mind, geographers must exert special care in their studies since the visual techniques used can be so persuasive that they can, at times, have more public impact than is methodologically justified. The heightened significance attributed to the GAM technique, for example, has resulted in it being described as a 'foolproof method of detecting clusters of child leukaemia' (Corke 1987), without any apparent awareness of the criticisms mentioned above.

Adjustment for confounders

When examining the spatial distribution of disease cases, it is extremely important to consider the effect of possible confounding variables (associated with both location and the particular disease) which may have a non-uniform geographical distribution. Clearly, if the distribution of known risk factors is non-random across space, then it will be reflected in the distribution of the disease itself. If these confounders are not taken account of in the analysis it may give the false impression of clustering or raised rates of disease which may then be attributed wrongly to a particular source of pollution; conversely, areas of raised incidence may be missed.

So, for example, if an area has a particular concentration of people of an age which is at greater risk from a disease than elsewhere, then we would expect to see a greater incidence there than in other areas. It would make sense, therefore, to standardize for age as a confounder by producing age-adjusted rates or stratifying by age. It is not unusual to see analyses performed and sometimes inferences drawn from such analyses using data unadjusted for obvious demographic factors like age or size of population at risk (Gatrell and Lovett 1989; Lovett *et al.* 1990: 106).

Dose response relationship

Another problem is that of detecting the effects of low doses. Since all epidemiological work must be based on detecting cases, either retrospectively or prospectively, it is likely that a high proportion of them will be related to higher doses and exposure levels. The problem then comes in deciding what dose-response model to assume when extrapolating to low doses (J. Smith 1988; Breslow and Day 1980). This has beset the debates concerning occupational safety levels for ionizing radiation, as much of our knowledge about the carcinogenic and teratogenic effects of radiation has been obtained from studies carried out on the survivors of the two atomic bombs dropped on Japan in 1945.

The main problem facing an analysis of the waste disposal sector relates to the scarcity of information about the health effects that is currently available to researchers. Many landfill sites, for example, will contain a cocktail of hazardous substances which may act synergistically to produce adverse health effects. Whilst data may be available for individual chemicals, the presence of a combination of substances and 'new' compounds formed within the dump site will generate significant problems for researchers.

Time factors

Another component of risk which needs to be taken account of in epidemiological analyses relates to the temporal dimension of exposure. The net effect may be cumulative and exposure to the source intermittent and possibly related, not only to the length of exposure, but also the age of people exposed. There may also be a long latency period between exposure and effect as is found in many cancers. These time-related factors can pose a serious problem for the interpretation of spatial analysis (for a detailed review of some of these issues see Gordis 1988).

In summary, then, we have seen that any attempt to identify risk associated with waste disposal, or any point source of pollution, is not easy and presents researchers with many technical and methodological problems. Even when studies have been carried out in a technically exemplary way there are still major problems which remain. There is the issue of multiple inferencing and mapping, and also the interaction between science and policy in which experts act as intermediaries and operate within the constraints imposed upon them by democratic principles. Given that many of the issues raised by waste disposal must be considered under conditions of great uncertainty, it could be argued that traditional statistical approaches are not so useful in this context. As a consequence, the remainder of this chapter advocates the adoption of a Bayesian approach to the analysis of such problems and considers its implications for public involvement in decision-making.

BAYESIAN APPROACHES

The hazardous waste issue is part of a complex agenda of environmental concerns which are currently in a state of 'scientific flux'. The conventional wisdoms operating within many of these issues (for example, global warming, the landfill of waste and high temperature incineration) have been subject to substantial challenges from both socio-political groups and also elements of the scientific community. The knowledge base associated with such environmental problems has been shown to be deficient and has required further research into the causes of, and 'cures' for, environmental degradation.

One of the objectives of scientific research is the attempt to collect information, thereby increasing the certainty of our knowledge about a phenomena, relationship or event. In the present context we want to ascertain the risk to individuals as a function of their proximity to a waste disposal site. How is this done? Classically researchers will investigate the problem by some method (some of which have been discussed above), collect data (usually by a form of sampling) and come up with an answer, a relative risk of (say) childhood leukaemia due to exposure from some form of waste.

One of the problems associated with this approach is that, generally, it only uses the sample data collected in that study to make an assessment of the risk. From these data, inferences are made about the 'true' level of risk. This is usually expressed in terms of a probability (risk) of death or disease, which is then perceived (by both scientists and politicians alike) as a scientifically determined and therefore accurate figure. However, by their very nature, risk assessments are bounded by uncertainty and limited by the problems identified above and the preconceptions and biases imposed on the analysis by the originators of the study. Another problem is that several studies can be simultaneously undertaken and each come to different and sometimes conflicting conclusions. Each analysis is somehow taken as being a *unique* and *separate* contribution to the research endeavour.

A clear (if rather extreme) statement of the basis of the classical approach is given by Openshaw *et al.* (1987) in their account of the statistical techniques used in the detection of clusters. They claim that their technique:

> comes with guarantees that it is totally unbiased with respect to all knowledge both known and as yet undiscovered ... it is totally objective ... Bias is excluded and prior knowledge rendered irrelevant because no selectivity is required.
>
> (Openshaw *et al.* 1987: 338)

This notion of a 'value-free science' has been extensively criticized in the literature (see Irvine *et al.* 1979) as has the role of expertise within the policy process (Collingridge and Reeve 1986). The Bayesian approach regards prior knowledge and belief as important components of the analysis even when deemed to be (in 'scientific terms') subjective. A Bayesian would start off an analysis by formulating prior beliefs about the risk associated with a given waste plant. These priors may be anything from vague and poorly-formed ones, to sharply-articulated judgements based on information and knowledge available before the study is carried out. Thus, for example, the prior belief in the geographical distribution of true cancer rates across areas (after adjustment for known factors) will be based on data presented in any previous studies, a knowledge of the areas and individual views as to whether there are *a priori* reasons why there should be differences, and so on.

There are two key features of the Bayesian approach.[1] These are first, that prior beliefs are incorporated explicitly into the inference process, and second, that prior probabilities often consist of the subjective assessment of events made by an individual. In classical approaches which attempt to be 'objective', such prior beliefs are disregarded. The degree of belief in levels of risk is considered to be dependent on the evidence derived from the current study only. This can lead to certain contradictions in the literature. It is also rather disingenuous in that the methodologies used, the variables examined, the interpretation of what might be spurious or real, and the believability of the results, are all a function of the prior beliefs and knowledge of the research team.

Bayesian approaches and the meaning of risk

What do we mean when we claim we have identified the risks associated with a given system or point source of pollution? Put simply, risk is essentially a probability of a given event or outcome (consequences) occurring. The subjective approach to probability rejects all conceptions that probability exists in the abstract (De Finetti 1976). Probability does not 'exist' independently of the evaluations we make of it. As a result there is little meaning in statements such as 'the probability of something occurring is *x*'. This goes to the heart of the old controversy between the objectivist and subjectivist conceptions of probability. Subjectivists, when examining a probability, consider issues such as the probability of what, in what circumstances and evaluated by whom? The assigning of a probability to an event is carried out in the context of the knowledge base and circumstances known to be relevant at the time. Thus our evaluation of a risk will vary according to the state of our knowledge and will subsequently be reviewed and enriched as more information becomes available. This process is formalized in Bayes theorem.[2]

On the other hand, the objectivist view of science sees probabilities as properties of the external world, and properties of objects which are inherent within them. This is a reification of the concept of probability, an apprehension of a product of human activity as if it were a fact of nature, obeying cosmic laws. Thus the probability of rain on a given day is seen somehow as 'in as the weather system'. This naturally corresponds to the classical (frequentist) approach to assessing probabilities by looking at long-run relative frequencies of events. A subjectivist view on probability interprets it as a personal assessment of an event happening. So rather than talking about the probability of something, as a subjectivist, I would refer to *my* probability of an event (Dowie 1989). A subjectivist thus views 'probability' as being the property of the person assigning the risk, whilst the objectivist sees it as the property of the object.

The two key features of the Bayesian perspective are thus the explicit incorporation of prior probabilities and the subjective assessment of these probabilities. It is argued in the rest of this chapter that the Bayesian approach is more useful for the arena of risk assessment and management, particularly as it relates to waste disposal.

MULTIPLE INFERENCING

One of the problems associated with an attempt to identify whether the exposure to hazardous waste disposal sites has any adverse effect on health is that of having to investigate many possible exposures to a variety of chemicals and several possible health outcomes. This is the basic 'multiple inference' problem which is faced by much epidemiological research and it can have a serious effect on our ability to establish causal relationships.

When examining a multiplicity of hypotheses, one has to adjust upwards the *p*-values obtained to take into account the likelihood of chance significant values (Thomas *et al.* 1985; Jones and Rushton 1982; Griffith *et al.* 1989b). So we are faced with a situation where, because so many associations are being examined, real issues, of public health interest, may be missed and ruled out as non-significant. In rare diseases, where only very high relative risks can show up as being statistically significant, these relationships are seen as being too strong to be biologically plausible, and so ascribed to some unknown confounder (Clayton 1989).

The problems of multiple inferencing have been described by Mack and Thomas (1985) who examined the risks of cancer in ninety sites in the body in two-hundred exposed census tracts for the two sexes separately. They considered some 36,000 associations in all, from which it is difficult to assess which associations were reported significant by chance and which were real.

When considering this problem from a public health or risk management perspective, we are more interested in the value of the rates found than the mapping of *p*-values. This is one of the fundamental problems which disease-mappers have always faced. Imagine we draw a map (see Figure 10.1), and for each area we calculate the expected number of cases (E_i) of a rare disease (assuming a Poisson distribution and using appropriate age and sex standardized rates). This allows us to obtain the observed number of cases (O_i) in each area (i) and calculate the ratio of the Observed to the Expected (O_i/E_i) values. (If the cases considered are deaths then this is known as the Standardized Mortality Ratio (SMR).) This ratio gives us an estimate of θ_i which is termed the relative risk.

As epidemiologists we are interested to see if any areas have ratios which are less than or greater than 1. If the ratio is greater than 1 then the rate of disease is higher than expected and may be worth examining further. We can either map these estimates of the rate or

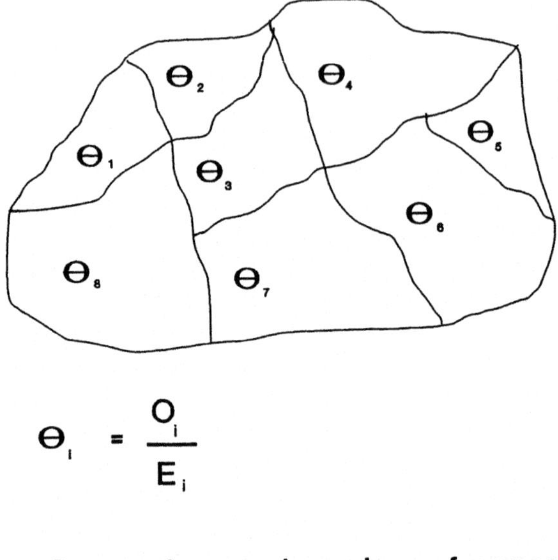

O_i = observed number of cases

E_i = expected number of cases

Figure 10.1 Distribution of relative risk

we can map the *p*-value (degree of statistical significance) of the estimate being different from 1.

One of the problems here is that those rates which are at the extreme ends of the distribution will tend to be so because of chance variation, that is, they will tend to be the areas which have very small expected counts and thus greater Poisson variability. Consequently, the map will tend to be dominated by areas which are seen as having a very high or low rate but where this is more likely to be due to chance. Of course, the reverse happens if the map is shaded according to *p*-values since a rate only really has a chance of being significantly different from the average in areas with large populations. So small relative risk changes will tend to dominate the picture and this might be of less importance from a public health perspective. Thus there is an inherent tension between mapping rates and mapping *p*-values (Muir 1981; Kaldor and Clayton 1987; Kemp *et al.* 1985).

One attempt to resolve this tension is by means of a compromise (see Clayton and Kaldor 1987; Kaldor and Clayton 1987; Thomas *et al.* 1985). This uses the empirical Bayes (EB) technique. Here, rather than using the calculated rate as the estimate for the true rate for each area, the empirical Bayes method pulls in the extreme rates towards the mean value for all the areas, a process referred to as shrinkage (see Figure 10.2). The degree to which each value is shrunk or pulled in towards the mean rate is related to the degree of variability of the estimate. The variability of the estimate itself is inversely related to the number of cases observed. The smaller the number of cases of the disease on which the rate is based, the more variable is the estimate, and so the more it is 'shrunk'. Conversely, the greater the number of cases on which the estimate is based, the more it is 'believed'.

The prior belief is that there is no spatial variation and this is then modified by the calculated rates. Thus, rather than letting the data from each area dominate the inference, some prior notion of the parameter value in an area is used to influence the estimate of θ_i. In the empirical Bayes technique, as opposed to a 'fully-blown' Bayesian approach, the prior information actually comes from the other areas (Clayton and Kaldor 1987). The EB estimate for an area is the weighted average of the estimated rate for that area and the average rate over *all* the areas. The weight, in this case, being determined by the number of cases and so the degree of precision of the estimates. Obviously adjustment is made for the age–sex structure of the population in each area so that the appropriate 'average' rate is used.

This basic model has a rather mechanical feel about it since the 'priors' here are suggested by the data rather than the researcher's subjective probability (Maritz and Lwin 1989). However, since the parameters used to weight the estimates of the rates are explicit, there is nothing to stop one changing the weighting of some of the areas if there are *a priori* reasons for doing so. For example, if it is believed that a particular area was somehow different from the 'global' pattern (having its own regional rate) then the average of this region could be used for weighting its districts instead of the national average. If it is supposed that there are reasons to assume greater homogeneity between adjacent regions, that is spatial autocorrelation, then the Conditional Autoregressive (CAR) process can be included. This is particularly the case where the rate in an area is both conditional on those over all and dependent on the

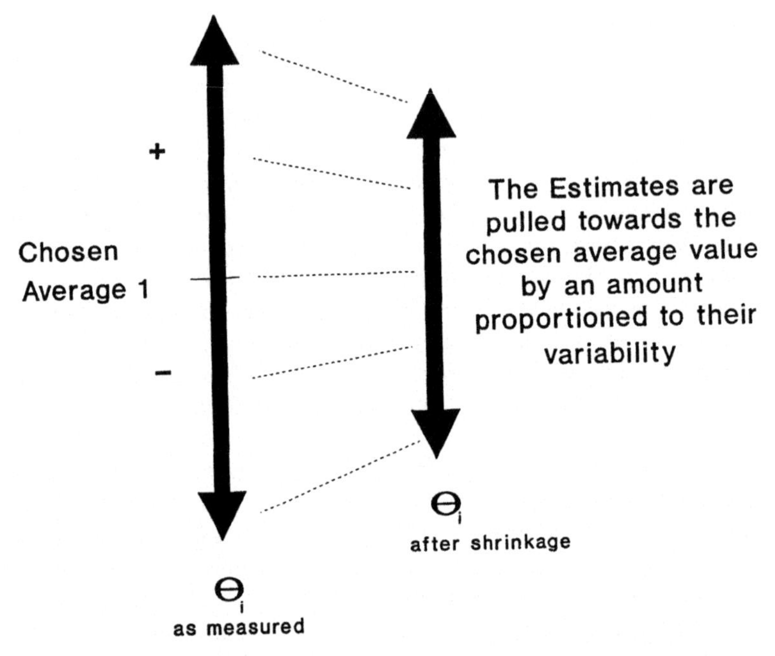

Figure 10.2 Shrinkage of estimates of relative risk by empirical Bayes method

average of adjacent areas (Besag 1974). The use of such a model therefore depends on the scale within which the geographical influences upon risk are varying relative to the size of the mapping areas. More controversially, given prior knowledge or beliefs about the conditions and presence of particular risk factors in one or more of the areas, the degree to which the estimates of the rates are 'pulled' towards the mean can also be changed. This allows a variety of groups to interpret the same data in a different way by making their differences (that is, prior assumptions) explicit. Thus, if one believes that the presence of a large toxic waste dump site is likely to raise the risk of certain diseases (perhaps because of established toxicological information) then the exposure of each district to the dump site (perhaps on the basis of distance) can be taken into account in the estimation of the relative risks.[3]

The sort of factors which could influence the prior distribution of the weights used can be anything from the biological plausibility of an association, through potential bias in design, the quality of the information on exposure, disease occurrence, previous work done, to 'local' knowledge. Clearly then, a geographical perspective, which brings to the study of spatial distribution more than just the physical distribution of risk factors, but a deeper understanding of the meaning of space and location, has a lot to offer in the development of spatial priors (Moon 1990).

This has demonstrated the potential of a Bayesian approach for the mapping of events and the investigation of multiple associations. These techniques have already been used to estimate the incidence of cancer in communities near toxic waste disposal sites (Thomas 1988). The process has been taken a step further by Clayton (see Cartwright *et al.* 1990, Appendix I) who has extended the Bayesian approach to mapping disease using a 'penalized likelihood' function which exerts a local smoothing to the relative risks. The new Leukaemia and Lymphoma map of England and Wales uses this approach and represents a major theoretical advance in disease mapping which needs to be incorporated into subsequent studies (Cartwright *et al.* 1990).

This procedure allows for a variety of rates to be presented for a given district depending on the shrinkage weights used. Consequently, it appears to strike at the heart of the objective 'scientific' approach. Not surprisingly, therefore, this is one of the reasons why there is great resistance to such methods by more traditional researchers. However, in its defence, it is no worse than other current practices such as the selective reporting of results and the choice of study areas because of the appearance of a cluster.

The improvement allowed for by a Bayesian approach is that the injection of the prior beliefs is both controlled and explicit. The more data there is then, the more conflicting posteriors will be drawn towards one common distribution. As such, 'rational' groups will no longer be able to hold extreme views with any degree of confidence, and this results in the formation of a consensus. The implications this has for policy-making and the democratic process will be discussed later.

DECIDING ABOUT RISK: THE POLITICS OF UNCERTAINTY

The estimates of risk found to be associated with a waste disposal site and the significance of these results do not, of course, have immediate implications for action. These need to be fed into a decision theoretic framework using a decision tree with cost-benefit values or utilities associated with alternative courses of action

(Lindley 1965: 62). Bayesian philosophy combined with this decision theoretic approach to inference helps in the choice between decisions. Taking this further, Thomas (1985) proposes that we should also consider the cost of both false positives and negatives within the framework of a decision tree. For example, rather than just considering the expected utility of following one decision path relative to another (where the risks associated with the paths are assumed to be the true ones), research should assess the degree of confidence associated with a risk estimate (in terms of significance levels or confidence limits). In this way the cost of reporting a false association (for example, a plant being closed with its associated economic costs) or a false negative (leading to the plant continuing but with resulting human suffering) is calculated. In this way the policy-makers can judge the full range of decisions that they may wish to make. This point has also been developed by Collingridge (1982, 1984) in terms of the error cost of decisions.

Thus, to take an extreme case, it is possible that the cost of underestimating the risk of a catastrophe is so great that it is better (in terms of potential costs) to make the assumption in the decision process that it *will* occur. There is evidence to suggest that public groups often behave in this way, for example, in the case of nuclear power, and have subsequently been criticized by corporate groups for their supposed irrationality. Alternatively, if the potential health effects associated with a site are so minor compared to the economic and social benefits it brings to a district, then decision-makers could weight their judgements accordingly. Whilst achieving consensus in contentious areas, such as waste disposal, is fraught with political difficulties, a more open Bayesian process would remove much of the potential for the exploitation of expertise which has plagued such debates in the past.

The Bayesian scheme of inference aids us in deciding which information and data sources are the most useful in discriminating between competing hypotheses since it looks at how new information changes our assessment of risk. In addition, it illustrates the implications of the collected information for hypothesis testing and also uses the resulting analysis to select the course of action that seems optimum in the light of the available evidence (Lindley 1972, 1985). However, care must be taken to avoid bias entering the analysis. This may occur in two ways. First, the information collected may not be given enough weight because of a feeling that one knows all along what the result is going to be. The second source of bias may occur if researchers give too much weight to any additional evidence thereby allowing it inappropriately to dominate the prior beliefs/information (Fischhoff and Beyth-Marom 1983). Both of these sources of bias have implications for policy-making and need to be highlighted if such work is going to have practical implications.

Policy-makers are normally forced to make decisions under conditions of uncertainty, the degree of which can be great where the scientific foundations of the available data themselves are unclear or where fundamental data are lacking. These decisions, though often made under conditions of ignorance, have important consequences in the case of environmental risks. Postponing decisions until more information is available can involve considerable economic costs and raises questions over the optimum or sufficient levels of information available to policy-makers. The relevance of a Bayesian analysis in this regard, with the priors carrying much of the weight, is obvious as is the notion of error cost detailed earlier (see Thomas 1985).

The assessment of health risks form only one strand of the environmental risk management process. Political decisions, pressure from various sections of the public and industry, and competing scientific evidence are also influential. These, in addition to the usual factors in decision-making such as the balance of costs, technical feasibility, availability of substitutes, are synthesized as part of the decision-making process. The relative influence of these factors varies between toxicants involved, sectors of the economy affected, along with sources and forms of waste disposal under consideration. A review of the background to the regulation of several chemicals and the role of these competing factors in US environmental policy is given by Needleman (1988) (also see Landrigan *et al.* 1989).

In contrast to the USA, the degree of public participation in the risk assessment process is generally very small in Britain and usually restricted to the public inquiry process. This is partly the result of a lack of resources and awareness, acting in combination with a predominant concern with issues of commercial confidentiality which conspires with the alienating/intimidating arena of science. However, within a democratic society public participation is deemed to be important. It ensures that governments and corporations do not have a monopoly over risk assessment and, as a result, cannot be allowed to dominate the environmental health agenda. It also allows those who bear the costs of an activity (that is, local publics) to have some say in determining its acceptability. The ability of community groups and local authorities to carry out their own environmental risk studies is an exciting prospect, although few groups have been successful in the context of UK debates. In th USA a movement towards such community epidemiological studies is developing and some epidemiologists see themselves acting as a resource for local groups (Legator *et al.* 1985). However, the experience of local groups in Britain using expertise to articulate their case has met with a mixed response in the past (as illustrated by the nuclear power debates and conflicts surrounding major hazards (see Smith 1990).

The Bayesian approach may facilitate this participation of social scientists and the public, for it allows us,

> to seek to report openly and accessibly, a rich range of the possible belief mappings induced by a given data set, the range being chosen to reflect and potentially to challenge the initial perceptions of a broad class of interested parties.
>
> (Smith 1986)

In this context it is perhaps worth questioning those who make a clear distinction between science and what Weinburg (1972) calls trans-science. Both Smith (1990) and Everest (1990) argue that the issues surrounding environmental risk are generally recognized as being trans-scientific, in that whilst amenable to the scientific approach they 'cannot be fully answered by science and so must incorporate an essential element of public and political judgement' (Everest 1990). Such trans-scientific problems involve a fusion of the scientific and social science approaches. To a degree this is an arbitrary division which assumes a degree of certainty in 'science' that is, in fact, illusory. In many ways the Bayesian approach unites activities across the scientific boundary in that it feeds and links into decision analyses based on alternative views. Bayesian decision theory therefore allows for a set of possible decisions to be evaluated in the light of new evidence and so does not view discord or disagreement as the product of a scientific deficiency.

A Bayesian scheme appears to be more relevant for policy decision-making where conflicting 'expert' opinion has to be aggregated and uncertainty prevails. Hodges (1987) discusses the various sources of uncertainty prevalent in policy analysis. The usual source of such uncertainty considered is that, having decided on a model for prediction of events, there is often a degree of ambiguity as to the parameters of that model (termed structural uncertainty). He stresses the importance of a framework to enable effective communication between members of a research team, an essential prerequisite of such communication being the acknowledgement as to the sources of uncertainty in the analysis and 'to incorporate them explicitly in choices made in the course of the analytical work and the products that arise from it' (Hodges 1987: 288). However, the reality of the risk communication process is often far removed from such an ideal.

When science lays claim to be 'objective', it is perpetuating a myth. Whittemore (1983) argues that the separation of science and policy is an erroneous description of reality and an unattainable goal for the regulatory process. She then goes on to explore the covert ways in which 'values' enter toxic risk assessment. First, scientists' own values determine both the quality and quantity of information used by researchers in the risk assessment. Second, risk-assessment procedures themselves involve value-laden procedures. For example, there are often a number of assumptions which underlie the various models used in the analysis. Third, individual values affect scientific interpretation of experimental results. Finally, she argues that values influence the weights used by researchers to combine disparate data into a coherent scientific argument as required by a multi-disciplinary risk assessment.

Nowhere are these points clearer than in policy-relevant scientific inquiry such as hazard identification, quantification and regulation (Ashford 1988). As a consequence, there are seen to be 'inherent tensions between the use of science in the policy process and democracy' (Dietz, quoted in Freudenburg 1989). Essentially, the regulators want a risk to be 'accurate' enough for action to be justified. However, the degree of truth and the basis of that truth required to legitimate policy will vary according to the circumstances and political dynamics of the debate. Anderson (1988) shows how different governmental agencies, operating within the same country, derive and use alternative estimates of a chemical's carcinogenic potency within policy formulation. There are many ways in which the research question can be framed because of the variety of potential data sources and methods of both analysis and presentation. Thus, what often appears to be disputes about science are, in reality, arguments over the values inherent in the underlying research assumptions (Rushefsky 1985).

The Bayesian approach, with its use of explicit priors incorporating knowledge that exists at the time of research, may help regulatory policy to integrate facts, judgements and values and go some way to the combining of conflicting judgements of fact and value.

There is not the space here, nor is this the proper place, for a *thorough* discussion of some of the general issues of the role of science in influencing and forming public policy. However, Hammond *et al.* (1983) detail the fundamental obstacles to the use of such 'scientific' information in the making of public policy. These can be summarized as follows. First, effective policy formation requires an integration of both 'facts' and social values, but there is rarely agreement from 'the experts' on the facts, and similarly no consensus of community values.

Second, scientific statements and, particularly, projections of the likely impact of alternative options are usually probabilistic with confidence intervals. This does not sit well with the immediate needs of policy-makers to decide on a singular, discrete choice.

Finally public policy-makers use scientific information (and indeed generate it) within a political context, thus politics may well influence the types of alternatives explored and presented in an assessment. (Hammond *et al.* 1983; for an entertaining review of extreme cases of these problems see Watts 1990).

Yet despite such strongly-articulated concerns, scientific data is still used *and* given considerable weight within policy areas such as hazardous waste disposal. Part of the reason lies in the supposed legitimacy of science and the power of those groups who use scientific knowledge for their own ends. Smith (1990) has explored the role of science in public sector decision-making for the location of major hazard sites. He raises the issue of the displacement of debates on risk away from the public arena towards the corporate sector, with its subsequent threat to the democratic process as a result of its increasing domination by scientific experts. He has also questioned whether the move towards scientific expertise in risk debates (through the increased use of risk analysis) has in some way taken us away from a pluralist view of environmental control by removing the power of the body politic to intervene effectively in the decision-making process.

It is here that the use of a subjectivist, Bayesian framework can pull together some of the strands. Not only does this approach allow for disagreements, it also makes them explicit, and thus helps to express scientific uncertainty. In addition, it makes clear that much of the scientific dispute on the assessment of risk is not so much a dispute about the studies performed but about certain prior ideas or values held by the expert analysts. However, since much of this is cloaked in very technical and statistical language, it effectively excludes the public from the debate.

CONCLUSIONS

> Bayes or empirical Bayes procedures promise some improvement over current procedures for estimation of a multiplicity of related effects ... and for synthesising expert opinion so as to more adequately express model uncertainty. It is perfectly appropriate in a democratic society that a carefully quantified measure of public opinion or belief be used to assist decision-makers with their task.
>
> (Breslow 1988)

It is often not clear how much influence scientific studies actually have in the policy process. This is partly because scientists have often hidden behind a mask of supposed objectivity attempting somehow to transcend the world of politics by laying claims to the neutrality of science. However, attempts by public groups to utilize science in their arguments have met with little success (Freudenburg 1989; Smith 1990). Indeed some writers have gone so far as to argue that,

> Instead the ironic implication is that the best way to achieve a more scientific outcome may be for scientists to pay greater attention to political factors, and specifically to increasing the access to scientific resources among the political parties involved.
>
> (Freudenburg 1989)

The ideas discussed here may be one step in this process. A move towards the adoption of a Bayesian framework would encourage a more explicit statement of the disagreements over hazard issues and also open the arena to a wider audience. Within such a framework risks can be assessed in terms of the priors held by local communities or groups and thus the origins of scientific conflict would become more transparent. This would allow for the removal of the 'cloak of commercial confidentiality' which has surrounded many UK hazard debates in the past. In addition, it would also have the potential to shift the balance of power from the corporate sector which currently is advantaged due to its superior financial and technocratic resource base, towards a more democratic decision process.

In this way the ideas of various sections of the public (for example, trade unions and local publics) will assume greater significance in the regulatory process and may thereby help to make and contest decisions in a manner that is regarded as truly scientific. Whilst the methodological problems of finding group utilities are substantial (J. Smith 1988; 73), as are the difficulties in getting consensus over the priors, the consideration of Bayesian perspectives could yield some interesting results. The importance of a Bayesian approach to the study of waste disposal by geographers and epidemiologists alike is apparent, and warrants further empirical research in this field to verify the methodology. Such research should aim to break down the barriers between disciplines, thereby allowing for a greater understanding of the nature of the problem.

ACKNOWLEDGEMENTS

Thanks go to David Clayton for so many stimulating ideas and to Keith Abrams for invaluable advice on the Bayesian philosophy.

NOTES

1 In the Bayesian approach the aim of research is to continuously add information so as to update or revise one's prior belief. The statistical basis of this approach is beyond the scope of this text. For a detailed account see J. Smith 1988; Ch. 6; Lindley 1965, 1972. For a more critical review of this approach see Cox and Hinckley 1974; Ch. 10.
2 For a detailed and comprehensive account of the Bayesian approach to scientific reasoning see Howson and Urbach 1989.
3 For a discussion of the possibility of weighting the districts according to prior information such as plausibility, existence of confounders and anatomical site of cancer, see Thomas 1988; Buffler *et al.* 1985.

REFERENCES

Aitkin, M., Anderson, D., Francis, B. and Hinde, J. (1989) *Statistical Modelling in Glim*, Oxford: Oxford University Press.

Anderson, H. (1985) 'Evolution of environmental epidemiologic risk assessment', *Environ. Health Persp.* 62, 389–92.

Anderson, P. (1988) 'Scientific origins of incompatibility in risk assessment', *Statistical Science* 3 (3), 320–7.

Ashford, N. (1988) 'Science and values in the regulatory process', *Statistical Science* 3 (3), 377–83.

Besag, J. (1974) 'Spatial interaction and the statistical analysis of lattice systems', *Journal of the Royal Statistical Society* Series B, 36, 192–236.

Breslow, N. (1988) 'Biostatisticians and Bayes', Paper to *Am. Stat. Assn.*

meeting, August 1989, Department of Biostatistics, Seattle: University of Washington.

Breslow, N. and Day, N. (1980) *Statistical Methods in Cancer Research*, 1, *The Analysis of Case-control Studies*, Lyon: International Agency for Research on Cancer.

Buffler, P., Crane, M. and Key, M. (1985) 'Possibilities of detecting health effects by studies of populations exposed to chemicals from waste disposal sites', *Environ. Health Perspect.* 62, 423–56.

Carstairs, V. and Lowe, M. (1986) 'Small area analysis: creating an area base for environmental monitoring and epidemiological analysis', *Community Medicine* 8 (1), 15–28.

Cartwright, R., Alexander, F., Ricketts, T. and Mackinley, P. (eds) (1990) *Leukaemias and Lymphomas: An atlas of distribution within areas of England and Wales 1984–1988*, Leukaemia Research Fund.

Clayton, D. and Kaldor, J. (1987) 'Empirical Bayes estimates of age-standardised relative risks for use in disease mapping', *Biometrics* 43, 671–81.

Clayton, D. (1988) 'The work of Openshaw, Craft, Charlton, Birch and Wymer, concerning cancer clustering in the north of England: a critical review', Report to the Northern Regional Health Authority.

Clayton, D. (1989) 'Invited discussion', *Journal of the Royal Statistical Society* Series A, 152 (3), 365–7.

Cliff, A. and Haggett, P. (1988) *Atlas of Disease Distributions, Analytic Approaches to Epidemiological Data*, Oxford: Basil Blackwell.

Collingridge, D. (1982) *Critical Decision Making*, London: Frances Pinter.

Collingridge, D. (1984) *Technology in the Policy Process – Controlling Nuclear Power*, London: Frances Pinter.

Collingridge, D. and Reeve, C. (1986) *Science Speaks to Power: the Role of Experts in Policy Making*, London: Frances Pinter.

Corke, R. (1987) 'Children in unsuspected areas could be at risk', *The Listener*, 26 November, 12.

Cox, D. and Hinkley, D. (1974) *Theoretical Statistics*, London: Chapman and Hall.

De Finetti, B. (1976) 'Probability: beware of falsifications', in A. Aykac and C. Brumat, (eds) (1977) *New Developments in the Applications of Bayesian Methods*, North-Holland.

Doll, R. (1955) 'Mortality from lung cancer in asbestos workers', *British Journal Ind. Med.* 12, 81.

Dowie, J. (1989) *Professional Judgement*, introductory texts 5–7 (for course D321), Milton Keynes, Open University.

Estes, J., McGwire, C., Fletcher, G. and Foresman, T. (1987) 'Coordinating hazardous waste management activities using geographical information systems', *Int. Journal Geog. Inf. Systems* 1 (4), 359–77.

Everest, D. (1990) 'The provision of expert advice to government on environmental matters: the role of advisory committees', *Sci. Publ. Affairs* 4, 17–40.

Fischhoff, B. and Beyth-Marom, R. (1983) 'Hypothesis evaluation from a Bayesian perspective', *Psychological Review* 90, 239–60.

Freudenburg, W. (1989) 'Social scientists' contributions to environmental management', *J. Social Issues* 45 (1), 133–52.

Gardner, M., Snee, M., Hall, A., Powell, C., Downes, S. and Terrell, J. (1990) 'Results of a case-control study of leukemia and lymphoma among young people near Sellafield nuclear plant in West Cumbria', *Brit. Med. Journal* 300, 423–9.

Gatrell, A. and Lovett, A. (1989) 'Burning questions: incineration of wastes and implications for human health', presentation at Institute of British Geographers' Annual Conference, Coventry, January.

Goldman, L., Paigen, B.J., Magnant, M. and Highland, J. (1985) 'Low birth weight, prematurity and birth defects in children living near the hazardous waste site, Love Canal', *Haz. Waste* 2, 209–23.

Gordis, L. (ed.) (1988) *Epidemiology and Health Risk Assessment*, Oxford: Oxford University Press.

Griffith, J., Duncan, R. and Hulka, B. (1989a) Biochemical and biological markers: implications for epidemiologic studies, *Arch. Environ. Health* 44 (6), 375–81.

Griffith, J., Duncan, R., Riggan, W. and Pellom, A. (1989b) 'Cancer mortality in US counties with hazardous waste sites and ground water pollution', *Arch. Environ. Health* 44 (2), 69–74.

Hammond, K., Mumpower, J., Dennis, R.L., Fitch, S. and Crumpacker, W. (1983) 'Fundamental obstacles to the use of scientific information in public policy making', *Technological Forecasting* 24, 287–97.

Hattis, D. (1988) 'The use of biological markers in risk assessment', *Statistical Science* 3 (3), 358–66.

Heath, C. (1983) 'Field epidemiologic studies of populations exposed to waste dumps', *Environ. Health Persp.* 48, 3–7.

Heath, C. (1988) Uses of epidemiologic information in pollution episode management, *Arch. Environ. Health* 43 (2), 75–80.

Hills, M. and Alexander, F. (1989) 'Statistical methods used in assessing the risk of disease near a source of possible environmental pollution: a review', *Journal of the Royal Statistical Society* Series A, 152 (3), 353–63.

Hodges, J. (1987) 'Uncertainty, policy analysis and statistics', *Statistical Science* 2 (3), 259–91.

Hogue, C. and Brewster, M. (1988) 'Developmental risks: epidemiologic advances in health assessment', in L. Gordis (ed.) *Epidemiology and Health Risk Assessment*, Oxford: Oxford University Press.

Howson, C. and Urbach, P. (1989) *Scientific Reasoning: The Bayesian approach*, Illinois: Open Court.

Irvine, J., Miles, I. and Evans, J. (1979) *Demystifying Social Statistics*, London: Pluto Press.

Irwin, A. (1985) *Risk and the Control of Technology*, Manchester: Manchester University Press.

Jones, D. and Rushton, L. (1982) 'Simultaneous inference in epidemiologic studies', *Int. Journal Epidemiol.* 11, 276–82.

Kaldor, J. and Clayton, D. (1987) *The Role of Advanced Statistical Techniques in Cancer Mapping*, presentation at Nineteenth International Symposium on Cancer Mapping, Dusseldorf.

Kemp, I., Boyle, P. Smans, M. and Muir, C. (eds) (1985) *Atlas of Cancer in Scotland 1975–1980. Incidence and Epidemiological Perspective*, IARC scientific publications, no. 72, Lyon, France.

Landrigan, P., Halper, L. and Silbergeld, E. (1989) 'Toxic air pollution across a state line: implications for the siting of resource recovery facilities', *Journal of Public Health Policy*, Autumn.

Legator, M., Harper, B. and Scott, M. (eds) (1985) 'A guide to the investigation of environmental health hazards by non-professionals', in *The Health Detectives Handbook*, Baltimore: Johns Hopkins University Press.

Levine, R. and Chitwood, D. (1985) 'Public health investigation of hazardous organic chemical waste disposal in the United States', *Environ. Health Perspec.* 62, 415–22.

Lilienfeld, A. and Lilienfeld, D. (1980) *Foundations of Epidemiology*, Oxford: Oxford University Press.

Lilienfeld, D. (1988) 'Changing research methods in environmental epidemiology, *Statistical Science* 3 (3), 275–80.

Lindley, D. (1965) *Introduction to Probability and Statistics from a Bayesian Point of View*, Part 2, *Inference*, Cambridge: Cambridge University Press.

Lindley, D. (1972) *Bayensian Statistics, a Review*, Philadelphia, SIAM

Lindley, D. (1985) *Making Decisions*, 2nd edn, New York: Willey.

Lovett, A., Gatrell, A., Bound, J., Harvey, P. and Whelan, A. (1990) 'Congenital malformations in the Fylde region of Lancashire, England 1957–1973', *Soc. Sci. Med.* 30 (1), 103–9.

Mack, T. and Thomas, D. (1985) 'Methodology for evaluating cancer risk in small communities: evaluation of cancer risk in the residential neighbourhood near BKK landfill', Report to the State of California Department of Health Services, University of Southern California, Dept of preventive medicine, June.

Maritz, J. and Lwin, T. (1989) *Empirical Bayes Methods*, 2nd edn, London: Chapman and Hall.

Moon, G. (1990) 'Conceptions of space and community in British health policy', *Soc. Sci. Med.* 30, 1.

Muir, C. (1981) *Report on the Workshop on Mapping of Cancer*, Lyon 10–11 December 1981, IARC Internal technical report 82/002, Geneva: World Health Organization.

Needleman, J. (1988) 'Sources and policy implications of uncertainty in risk assessment', *Statistical Science* 3 (3), 328–38.

Nelkin, D (1975) 'The political impact of technical expertise', *Social Studies of Science* 5, 35–54.

Openshaw, S., Charlton, M., Craft, A. and Birch, J. (1988) 'Investigation of leukemia clusters by use of a geographical analysis machine', *Lancet* i, 272–3.

Openshaw, S., Charlton, M., Wymer, C. and Craft, A. (1987) 'A Mark 1 Geographical Analysis Machine for the automated analysis of point data sets, *International Journal of Geographical Information Systems* 1 (4), 335–58.

Rothenberg, R., Steinberg, K. and Thacker, S. (1990) 'The public health importance of clusters', *Amer. Jour. Epid.* 132 (1) (Supplement), 3–6.

Rushefsky, M. (1985) 'Assuming the conclusions: risk assessment in the development of cancer policy', *Politics and the Life Sciences* 4, 31–44.

Smith, A. (1984) 'Present position and potential developments: some personal views. Bayesian statistics', *Journal of the Royal Statistical Society* Series A, 147, 245–59.

Smith, A. (1986) in discussion of B. Efron (1986) 'Why isn't everyone a Bayesian?' *The Amer. Statist.* 40, 1–11.
Smith, A.H. (1988) 'Epidemiologic input to environmental risk assessment', *Arch. Environ. Health* 43 (2), 124–7.
Smith, D. (1990) 'Corporate power and the politics of uncertainty: conflicts surrounding major hazard plants at Canvey Island', *Industrial Crisis Quarterly* 4 (1), 1–26.
Smith, J. (1988) *Decision Analysis, a Bayesian Approach*, London: Chapman and Hall.
Spitzer, W., Shenker, S. and Hill, G. (1982) 'Cancer in a Montreal suburb: the identification of a nonepidemic', *Can. Med. Assoc. J.* 127, 971–4.
Stallones, R. (1988) 'Epidemiology and environmental hazards', in L. Gordis (ed.) *Epidemiology and Health Risk Assessment*, Oxford: Oxford University Press.
Stern, A., Munshi, A. and Goodman, A. (1989) 'Potential exposure levels and health effects of neighbourhood exposure to a municipal incinerator bottom ash landfill', *Arch. Environ. Health* 44 (1), 40–7.
Stubbs, J. (1989) *Cuba: The Test of Time*, London: Latin American Bureau.
Thomas, D. (1985) 'The problem of multiple inference in identifying point-source environmental hazards', *Environ. Health Perspect.* 62, 407–14.
Thomas, D. (1988) 'Development of a methodology to investigate cancer incidence in communities near toxic waste disposal sites', Final report to State of California Department of Health Services, Department of Preventive Medicine, University of Southern California, Los Angeles, June.
Thomas, D., Siemictycki, J., Dewar, R., Robins, J., Goldberg, M. and Armstrong, B. (1985) 'The problem of multiple inference in studies designed to generate hypotheses', *Amer. Journal of Epidemiol.* 122 (6) 1080–95.
Twigg, L. (1990) 'Health-based geographical information systems: their potential examined in the light of existing data sources', *Soc. Sci. Med.* 30 (1), 143–53.
Vianna, N. and Polan, A. (1984) 'Incidence of low birthweight among Love Canal residents', *Science* 226, 1217–19.
Watts, S. (1990) 'Science advice: an abuser's guide', *New Scientist* 125, 1707, 10 March.
Weinburg, A. (1972) 'Science and trans-science', *Minerva*, 10, 209–22.
Whittemore, A. (1983) 'Facts and values in risk analysis for environmental toxicants', *Risk Anal.* 3, 23–33.
Wynne, B. (1982) 'Nuclear decision making – rationality or ritual?' *British Society for the History of Science*, London.

Chapter 11

Licensed to dump? A report on British Coal's sea dumping in Durham

Jonathan Renouf

INTRODUCTION

> This desecrated landscape is probably the worst example of wasted coastline in England.
>
> (*New Scientist*, 1981, quoted in Durham County Council 1983)

The state of the Durham coastline from Seaham to Crimdon is acknowledged to be a national disgrace. British Coal (BC) dumps in excess of three million tonnes of liquid and solid waste, which amounts to 90 per cent of the total amount of industrial waste dumped into the North Sea. The result is an environmental disaster area. However, British Coal has claimed that it cannot afford to pursue alternative inland dumping options without jeopardizing the economic future of the six remaining Durham collieries. Appeals to the government to fund the cost of inland dumping have been turned down on the grounds of the government's commitment to the 'polluter pays' principle. This chapter argues that the government's refusal to fund alternative dumping rests far more on questions of political expediency than it does on principle.

The government's treatment of the dumping problem seriously calls into question their claim to be committed to 'green' politics. For several years the government has manoeuvred to avoid taking action to stop beach dumping. They have failed to monitor and enforce their own licences and by-laws, and tried to play down the environmental impact of colliery spoil. On occasions they have misled protestors seeking to stop the practice.

In order to examine and evaluate these allegations, this chapter is split into four sections. The first examines the history of dumping and the debate over its environmental impact. The next considers the various regulatory mechanisms relevant to coastal dumping, and the way in which they have been used. Third, the various arguments for and against sea dumping are discussed, along with some of the alternative options available for waste disposal. Finally, a concluding section argues that only government funding of inland dumping can achieve the twin objectives of saving jobs and safeguarding the environment.

Before proceeding further, it should be noted that since the research for this chapter was completed in November 1988, there have been several new developments. I have tried to incorporate these by updating the text. However, this chapter should be read as an historical account since publishing deadlines mean that events will have superseded this text even before it reaches the printers.

HISTORY AND ENVIRONMENTAL EFFECTS

Two types of waste are produced from the coal production process. *Solid waste* consists of shale and other stone separated from coal in the washing process. *Liquid waste* constitutes the fluid used in the washeries to separate and clean the coal. The environmental effects of waste disposal are intimately related to the historical development of the practice, which is why these two aspects of the problem are considered together.

Solid waste

Colliery waste has been dumped in the North Sea since before the last war. To begin with, amounts were relatively small, as mining technology before mechanization involved bringing very little stone to the surface. Coal and stone were kept separate underground, and stone used as packing and support material. However, with the advent of 'total extraction' mining, all rock and coal were brought to the surface together and separated in coal washeries. Coastal collieries found that the easiest and cheapest method of disposing of these increasing quantities of solid waste was to tip them into the sea.

To begin with, small amounts of waste were removed by the natural action of the sea. As quantities increased, waves were unable to remove all the waste being dumped. Gradually the waste spread out over the foreshore and below the low water mark. As early as

Table 11.1 Dumping of solid waste in the North Sea

Licence period	*Licence quantity (tonnes per annum)*	*Amount dumped (tonnes per annum)*
1/6/74–31/5/75	2,575,500	1,797,294
1/6/75–31/5/76	2,570,400	1,076,485
1/6/76–31/5/77	2,570,400	1,545,190
1/6/77–31/5/78	2,500,000	1,470,518
1/6/78–31/5/79	2,500,000	2,203,464
1/6/79–31/5/80	2,500,000	2,208,588
1/6/80–31/5/81 [1]	2,500,000	2,418,905
1/6/81–31/5/82	2,500,000	2,360,453
1/6/82–31/5/83	2,500,000	2,631,937
17/6/83–30/6/83	205,000	
1/7/83–31/8/83	410,000	2,173,242
1/9/83–31/5/84 [2]	1,885,000	
1/6/84–31/5/85 [2]	2,500,000	188,463
1/6/85–31/5/86 [2]	2,500,000	812,896 [3]

Source: Letter from MAFF (John Selwyn Gummer) to Jack Dormand, MP for Easington, 24 March 1986.

Notes
1 Five-year licence.
2 Five-year licence expiring 31 May 1988.
3 To December 1985.

1956 Lord Shinwell raised the problem in Parliament. From then on the situation has steadily deteriorated, and a succession of government reports, visits and inquiries have all failed to change the situation.

Dumping continued to increase in the 1970s and early 1980s (see Table 11.1). Most significantly, however, tipping became concentrated at just two sites: Dawdon in the North, and Easington in the South (see Figure 11.1). Quantities previously spread along the coast were now piled up at just two sites. Most of the increase has taken place at Dawdon. The Easington disposal point deals mainly with waste from Easington Colliery.

Most solid waste dumped at Dawdon is processed by a second, privately-operated, washery at Bankside, owned by the Seaham Harbour Company. Prior to 1987 the washery was situated at Noses Point (where it operated without a planning licence before 1983). The washery takes the waste (now delivered by a British Coal conveyor to the Bankside site), and processes it a second time to

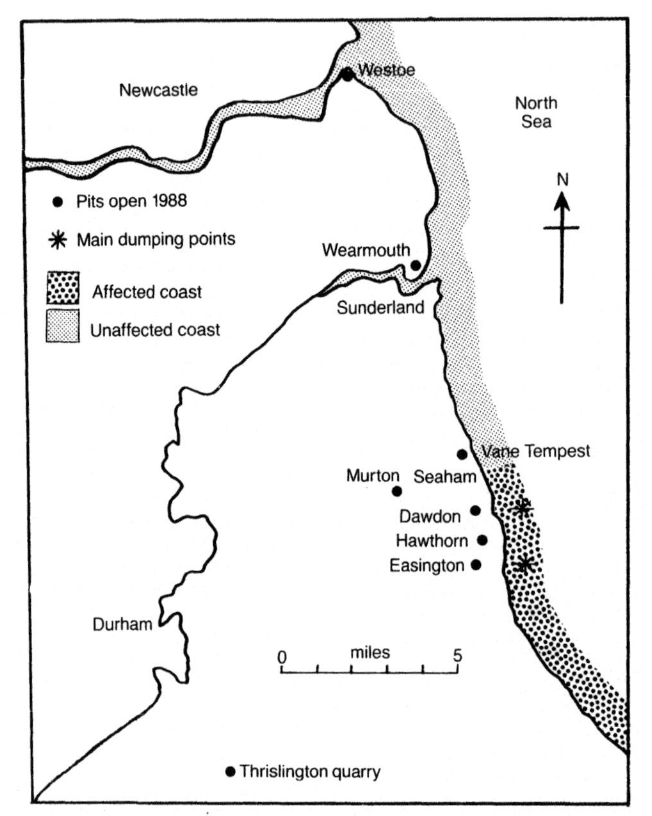

Figure 11.1 Dumping colliery waste in the North Sea

extract coal left following the initial BC washing, before finally bulldozing it over the cliffs into the sea.

The effects of dumping solid waste have been dramatic. Most obviously, the amenity value of the beaches has been destroyed. Beaches along a seven-mile stretch of coastline resemble black and churned up deserts. The sea itself is black. Spoil has spread out to smother the shore and the underwater environment. It kills the seaweed and other life on the sea-bed. Destruction of this habitat kills organisms which depend on sea-bed life for their food. The effect on the fishing industry has been significant, as the Ministry of Agriculture, Fisheries and Food (MAFF) report dealing with the disposal of solid wastes, including power-station fly-ash and colliery spoil, concluded:

Dumping has caused direct interference with commercial trawling and potting in the primary area of tipping.

(MAFF 1979: 33)

Liquid waste (or 'tailings')

The dumping of liquid waste is more recent than the dumping of solids. It has grown up because of new coal-washing techniques which involve washing coal with detergents to separate out impurities. Quantities have built up rapidly since the early 1970s.

Liquid waste has introduced huge amounts of suspended particulate matter into coastal waters. The Northumberland Water Authority estimate that about half a million tonnes of suspended solids are included in washery waste. Suspended matter irritates filtering creatures and blocks up gills, leading to large black residues forming inside crabs, and killing many crustaceans. Furthermore, the impact of washery detergents on marine life has yet to be fully investigated.

Combined effects

Despite these arguments, MAFF have consistently tried to play down the impact of colliery spoil on the coastal ecology. It has been left to protesting fishermen to gather samples from the seabed, and submit them for analysis by the Ministry. In their report MAFF deny that the samples contain unusual levels of colliery spoil, thereby suggesting that dumping affected a far smaller area than protestors claim. Nevertheless, drivers collecting samples reported that the seabed was smothered in a thick layer of sludge, with no marine life apparent.

So lax has MAFF's monitoring of the dumping operation been that it is currently the subject of an Ombudsman's enquiry.

In the absence of a detailed survey, it is difficult to ascertain the extent to which the sea-bed has become covered by colliery spoil. But while Greenpeace have yet to produce a video showing a piece of rock toppling off the end of a BC conveyor and killing a fish as it enters the water, there is little argument that where waste *does* cover the sea-bed, the impact on marine life is severe. In a survey conducted before the coal strike in 1984, Dr Nelson Smith from the Marine Biology Department at University College, Swansea, discovered that marine life on the affected beaches had been severely reduced. At Noses Point, adjacent to the main Seaham dumping sites, only *four* species of plant and animal life were found. A similar but unpolluted stretch of Northumberland coast yielded *eighty-four* species (letter from Dr Nelson Smith to William Thompson, 17 March 1988).

Commenting on the beach, he said:

> We have not investigated the biota of the major part of the foreshore (that is, the beach of sorted colliery spoil) for the simple reason that it appears to support no macroscopic life. I cannot attest to the influence of spoil tipping upon the original beach infauna because I am unable to determine the nature of the original beach.
>
> (ibid.)

There seems little doubt that areas affected by the waste are severely affected by it. The only argument concerns the extent to which waste coverage has spread beyond the immediate vicinity of the dumping sites. This dispute assumes further importance in the following section, which considers the various regulatory mechanisms for controlling dumping.

REGULATION OF WASTES

One of the most disturbing aspects of North Sea dumping has been the attitude of the responsible authorities to the regulation of waste disposal. No government department or regional authority has interpreted laws, which were clearly intended to limit pollution, in the spirit in which they were intended. Indeed, they have sought to evade their responsibilities, and turned a blind eye to law-breaking. This can be seen both with respect to solid and liquid wastes.

Solid wastes

Since the passage of the Dumping at Sea Act 1974, MAFF have been the authority responsible for licensing the dumping of solid wastes in the North Sea. As Table 11.1 shows, licensed quantities remained steady throughout the period June 1974 to May 1986 at around 2.5 million tonnes per annum. More recent licences show a reduction to just over 1.5 million tonnes. MAFF's justification for licensing colliery spoil disposal on Durham beaches rests on two claims. First, they claim that by 1974 the Durham beaches were already seriously damaged, and further dumping would therefore cause no additional damage to the environment. Second, dumping prior to 1974 had not been subject to any statutory controls. Hence R.A. Hathaway (Fisheries Officer at MAFF) wrote to William Thompson in March 1987 saying

> The tipping began 30 or more years ago, well before the advent of statutory controls over the disposal of waste at sea ... to the best of our knowledge the situation has not significantly deteriorated since 1974.

However, both these claims are difficult to sustain. Whilst it is true that total dumped quantities have not increased significantly over the past few years, it is not necessarily true that no extra damage has been caused. One reason for this is that whilst in the 1970s dumping was spread along the coast, it has now become concentrated at just two principal sites, Seaham and Easington. So although the total amount dumped on Durham beaches has not increased significantly, the amount being dumped at the two main dumping areas *has* increased, as quantities which were disposed of further down the coast have been diverted north. In fact, most of the increase has taken place at Seaham, because Easington is only equipped to deal with waste from its own colliery. Whilst there has been some improvement at beaches where dumping has ceased due to pit closures, it seems not unreasonable to assume that the increase in dumping at Seaham will have caused additional damage as these larger quantities spread out over a bigger area.

MAFF's claim that prior to 1974 dumping was not subject to statutory control is also dubious. Under the Sea Fisheries Regulation Act 1966, each Sea Fisheries District had the power to set its own by-laws, which were then confirmed by MAFF. By-law xv of the North East Sea Fisheries Committee was headed 'detrimental substances: deposit of'. It said simply: 'The deposit or discharge of any solid or liquid substance detrimental to sea fish or sea fishing is hereby prohibited'.

Since this by-law was in force before 1974, it seems clear that dumping *was* subject to legal control. And it would be difficult to argue that colliery waste was not a detrimental substance, as defined by this by-law. So whilst there may not have been a statutory *licensing* requirement prior to 1974, MAFF had confirmed by-laws which clearly prohibited dumping by the National Coal Board

(NCB). After 1974 they defied those by-laws by licensing dumpings which broke them. Not until 1989 did the Water Act rescind the power of the Sea Fisheries Committees to enact by-laws relating to detrimental substances.

Breaking the MAFF licence

Table 11.1 shows that for the year 1982–3, British Coal dumped 131,937 tonnes more than the MAFF licence allowed for that year. However, this relatively minor infringement does not include the quantities of solids which were discharged as suspended matter in liquid wastes. Even if we accept that from 1986 the Northumberland Water Authority (NWA) took responsibility for liquid discharges (see pp. 196–8), it is clear that *before 1986 the MAFF licence included liquid wastes.*

According to the NWA,

> assuming an overall average of 25% dry solids content . . . then the total dry solids in the discharges of coal washery effluents at Dawdon and Easington are about half a million tonnes a year. There are so many variables that it is not worthwhile to try to obtain an answer that looks more precise.
>
> (Letter from Sir Michael Straker, Chairman NWA, to John Cummings, MP for Easington, 24 March 1988)

If it is assumed that half a million tonnes of suspended solids were contained within the liquid wastes deposited every year, then Table 11.2 shows that between 1978 and 1984 the NCB regularly broke the terms of the MAFF licence.

When protestors raised the discrepancy with MAFF, the Ministry attempted to deny that they were the responsible authority. In a letter dated 17 March 1986 to Jack Dormand (then MP for Easington), John Selwyn Gummer (then the Minister at MAFF), argued that:

> Both the Dumping at Sea Act 1974 and Part 2 of the Food and Environment Act 1985 define 'sea' as 'including any area submerged at mean high water springs'. My officials are satisfied from their own observations on the spot, and from charts that, whatever the situation in the past, the mean high water springs at Seaham are currently well below the end of the pipe from which the tailings are discharged.

Table 11.2 Dumping of solid waste in the North Sea including suspended solids

Licence period[1]	*Licence quantity (tonnes per annum)*	*Amount dumped (tonnes per annum)*	*Excess (tonnes per annum)*
1/6/74–31/5/75	2,575,500	2,297,294	−278,206
1/6/75–31/5/76	2,570,400	1,576,485	−993,915
1/6/76–31/5/77	2,570,400	2,045,190	−525,210
1/6/77–31/5/78	2,500,000	1,970,518	−529,482
1/6/78–31/5/79 } 2	2,500,000	2,703,464	+203,464
1/6/79–31/5/80 } 2	2,500,000	2,708,588	+208,588
1/6/80–31/5/81 } 2	2,500,000	2,918,905	+418,905
1/6/81–31/5/82 } 2	2,500,000	2,860,453	+360,453
1/6/82–31/5/83 } 2	2,500,000	3,131,937	+631,937
17/6/83–31/5/84	2,500,000	2,673,242	+173,242
1/6/84–31/5/85 } 3	2,500,000	188,463	4
1/6/85–31/5/86 } 3	2,500,000	812,896	4

Sources: Letter from MAFF (John Selwyn Gummer) to Jack Dormand (then MP for Easington) 24 March 1986, and letter from Sir Michael Straker (Chairman NWA) to John Cummings (MP for Easington), 24 March 1988.

Notes
1 After 31 May 1988 MAFF granted a temporary one-year extension of the dumping licence.
2 Five-year licence.
3 Five-year licence expiring 31 May 1988.
4 Figures affected by strike and strike recovery period. I have not added on any figure for suspended solids.

On this basis, the Minister disclaimed responsibility for liquid waste. Some of the logic of this argument was truly Orwellian, since the only reason the pipes were above high water mark was because of the effect of dumping in raising the beach level!

However, the MAFF argument soon came under attack. In a letter to Councillor Ramshaw, on 29 April 1986, James Wilson (Chief Planning Officer for Durham County Council) reported:

> I have checked with the NCB today and am informed by the Area Minerals Manager that the licence issued by the MAFF under Section 2 of the Dumping at Sea Act, which extends to 31st May 1988, covers colliery waste, *including tailings.* [emphasis added]

Furthermore, the NCB's view is confirmed by the MAFF licence itself. Under Part 3 of the licence, the substances covered are described as 'colliery waste and liquid tailings'.

Therefore, until 1985 the MAFF licence included liquid waste but the NCB figures for the amount dumped did not include the suspended solids within the liquid waste. When these solids are added on to the NCB figures, they show that the NCB broke the MAFF licence with impunity for six consecutive years. However, help was at hand in the shape of the 1985 enactment of Part 2 of the Control of Pollution Act 1974 (CPA).

Liquid wastes and the Northumberland Water Authority

Only in 1986 did MAFF realize that section 32 of the Control of Pollution Act gave responsibility for licensing discharges of liquid wastes into coastal waters to the relevant regional water authority. This was outlined in a letter to William Thompson (Secretary of the Seaham Boat Owners Association) from R.A. Hathaway (Fisheries Officer, MAFF) on 31 May 1986:

> My belief is that such (pipeline) discharges fall within the scope of Part 2 of the Control of Pollution Act 1974, and are therefore the responsibility of Water Authorities.

Since Part 2 of the CPA was not enacted until 1985, MAFF and the NWA argued that prior to 1985, there were no legal restraints on liquid dumping. However, this ignores the terms of the MAFF licence, which included liquid wastes up until 1986. Furthermore, it also ignores North East Sea Fisheries Committee by-law xv, banning the dumping of 'detrimental substances'.

With the enactment of Part 2 of the CPA, the NCB was required to apply for a 'deemed consent' from the NWA to continue discharges. The position was explained by R.S. Stead (principal water quality officer, NWA) in a letter to William Thompson on 25 June 1987.

> This authority is responsible – under the Control of Pollution Act 1974 – for consenting discharges of effluents. Existing discharges of effluents have not so far required consent, but must be the subject of an application under the 1974 Act before 15th October 1987. Once a valid application has been made, the discharge has a deemed consent, which is unconditional, until the authority either refuses consent or issues a conditional consent.

In other words, permission is automatically given in the first instance, but after October 1987 the Water Authority must consider each consent, and either refuse it, modify the terms, or allow it to continue unchanged.

Unfortunately, despite an apparent desire to halt slurry discharges, the NWA has made clear that it feels that there is no point in opposing BC's application. On BBC Radio 4's *Face to Face* programme, Mr Stead (the NWA officer responsible for the licensing) argued that it would be futile to resist the application, because British Coal would simply appeal to the government, and the Water Authority would be overruled (7 October 1987).

The most controversial slurry consent is the tanker disposal point, known locally as the 'pig trough', at Dawdon. This is licensed to dump 38 litres/second, 24 hours a day, 5 days a week (see Table 11.3), which works out at around 180 large (18 cubic metre) tankers every day. However, several counts by local residents have shown tanker numbers around double this number. Even allowing for different size tankers, a recent count (over 2 and 3 March 1988) revealed that 87 per cent more slurry had been brought in for that 24-hour period than allowed under the consent. British Coal have argued in response that not all the tankers are full when they are brought to the disposal point – an argument lacking in credibility on purely commercial grounds. The tanker disposal point is the only dumping activity amenable to any sort of public monitoring, and every investigation has suggested that the terms of the deemed consent are regularly being flouted. It appears that the NWA have simply accepted BC's word with regard to the amount being

Table 11.3 Dumping of liquid waste in the North Sea

Disposal point	*Licensed quantity*[1]
Easington	23 litres/second (l/s)
Dawdon	28 l/s[2]
Tanker disposal point	38 l/s
Bankside	1,100 cubic metres

Source: Northumbrian Water Authority.

Notes

1 Twenty-four hours a day, 5 days a week.

2 Twenty hours a day, 5 days a week.

discharged. Their apparent failure to adequately monitor and control BC's dumping operation has led to an investigation of the NWA by the local Ombudsman.

Seaham Harbour Dock Company

Clause 37 of the Seaham Harbour Dock Company (SHDC) by-laws states:

> No person shall deposit or throw into the waters of the harbour any rubbish *or other material whatsoever* or place it in such a position that it is likely to fall, blow or drift into the harbour. [emphasis added]

The map showing the harbour limits encompasses the Bankside washery plant operated by the Dock Company. The waste from the washery is bulldozed into the sea as part of the solid waste on the 2.5m tpa MAFF licence. SHDC are operating the washery in defiance of their own by-laws.

New regulations

On 24 and 25 November 1987, the Environment Minister, Nicholas Ridley, chaired the Second International Conference on the Protection of the North Sea. A unanimous declaration was made following the meeting, and on 3 February 1988 the Minister issued a news release and guidance note explaining the government's interpretation of the declaration.

The intention of the declaration was made clear in paragraph 18 of the guidance note (paragraph 21a and b of the declaration):

> North Sea States have accepted as matters of principle that:
> (a) the dumping of polluting materials in the North Sea should be ended at the earliest practical date; and
> (b) *as from 1 January 1989, no material shall be dumped in the North Sea* unless there are no *practical* alternatives on land and it can be shown to the competent international organisations that the materials pose no risk to the marine environment. [emphasis added]

In its guidance note explaining the declaration the government offered this interpretation of paragraph 22a:

> Colliery spoil disposed of by the UK into the North Sea and, in some cases, on adjacent beaches, falls into the category of 'inert materials of natural origin', *which can still be disposed of at sea* (para 22 (a) of Declaration). Nevertheless, efforts will continue to be made to find environmentally acceptable alternatives to the current means of disposal of these wastes which would not jeopardise the economic future of the pits concerned. [emphasis added].

Although it is the clear intention of the declaration to stop the deposit of all materials except those which 'can be shown to the competent international organisations that the materials pose no risk to the marine environment', the government has manoeuvred to interpret the declaration to try and exclude colliery spoil. However, even if the argument is accepted with regard to solid waste, it cannot be claimed that liquid waste – which contains detergents used in the washing process – falls into the category of 'inert materials of natural origin'.

Application for increased discharge

At the beginning of 1988 British Coal applied for a new consent to increase discharges from the tanker disposal point ('pig trough') at Dawdon by twenty-five tankers a day. They argued that the slurry disposal point for Wearmouth and Westoe collieries (the former Boldon colliery shaft) is filling up, and they wanted to extend its life while looking for alternative options. However, there were suspicions that British Coal was preparing to divert *all* its slurry from Westoe and Wearmouth – which would involve 800 tankers a week.

This impression is supported by reference to a discussion document produced by the NCB in 1985 dealing with shale and fines disposal (washery tailings) at Dawdon colliery, which costed the operation for 20 years, and which *included estimates for 200 tankers a day from Westoe.* British Coal have therefore been considering the use of Dawdon as a liquid waste disposal point for Westhoe for three years. Their argument that they need more time to consider alternative sites rings hollow in the light of this discussion document. There has been plenty of time to consider alternatives.

In their application, British Coal failed to answer many important questions about their slurry dumping proposal for Dawdon. They did not produce detailed costings of all other options, including Hawthorn and Thrislington quarries. They did not say where they are considering dumping the remaining liquid waste that will come from Wearmouth and Westoe after the Boldon shaft is full. They did not prove that dumping at Dawdon is the cheapest option, or how collieries elsewhere in Britain are able to survive without a nearby beach.

Under pressure from massive local protest – including a formal objection from Durham County Council – the NWA rejected BC's application for an increased discharge. On 14 April 1988 British Coal announced their response by seeking an increase in discharge from the Easington disposal point by 727 cubic metres a day from 2,333 cubic metres (23 litres a second) to 3,060 cubic metres. This application was also turned down. Finally, British Coal was authorized to barge liquid waste from Westoe and Wearmouth out to sea for disposal.

IS THERE AN ALTERNATIVE?

Arguments for continued dumping

Why then has the government appeared so reluctant to act to end coastal dumping? Since 1988 the newest incarnation of Thatcherism has proclaimed itself committed to environmental issues. However, the evidence from this example reveals that, as far as Durham is concerned, the reality behind the political posturing is of a government committed to doing nothing to end coastal pollution. What then are the arguments used to justify continued dumping?

British Coal argue that whilst dumping is environmentally unattractive, it is a necessary evil because there exists no viable alternative which would not threaten the future of the pits concerned. All other options would increase the cost of dumping, thereby increasing operating costs, and therefore jeopardizing the future of North-East pits. This is a powerful argument, and its deployment has invariably silenced critics of coastal dumping. No one in Durham wants to see the remaining 8,000 coal-mining jobs lost. If it is a choice between jobs or clean beaches, jobs win every time. Neither is this situation altered by the recently-achieved profitability of Durham collieries. Their economic situation remains precarious, particularly in respect of the privatization of the Central Electricity Generating Board (CEGB) which will expose Durham coal to intense competition in its main Thames market (McCloskey and Prior 1988).

Attention has therefore shifted to the government, which has been asked to supply funds to pay for the additional cost of inland dumping. In 1985 a consultancy report by Ove Arup and partners, commissioned by the Department of the Environment, estimated that it would cost £10 million a year to send the waste to Thrislington quarry near West Cornforth. In 1988 British Coal commissioned another consultancy report which estimated the total cost of dumping waste from Dawdon and Seaham collieries at Thrislington as £44 million over ten years, and £60 million over twenty years for Easington colliery (Durham County Council 1989). Up until the end of 1989 the government had refused to finance this (or any other) option(s). In a Parliamentary answer in July 1986, the Junior Minister for the Environment, Mr R. Tracey, explained government philosophy:

> The disposal of waste to quarries inland would present fewer environmental problems, but the site proposed [presumably Thrislington quarry] does not have sufficient capacity to take all the spoil that is expected to be produced over the life of the collieries. *The costs of transportation are in any event high and could not be borne by the collieries concerned without placing their futures in jeopardy.*
>
> We have carefully considered whether these additional costs should be funded by central government. *We have concluded that this would not be in accordance with this government's commitment to the polluter pays principle.* It must be for the industry – and ultimately the consumer – to pay the costs of meeting the environmental standards of the day. [emphasis added]

The government therefore has attempted to avoid funding alternative dumping options by appealing to a supposedly universal principle – the polluter should pay. However, deeper examination of the issue reveals inconsistencies in the deployment of this 'universal' principle. Indeed, its use seems to relate more to questions of political choice than philosophical consistency. This can be seen most clearly when the treatment of colliery spoil is compared with the treatment of nuclear waste.

In December 1988 detailed plans for the privatization of the electricity generation and supply industries were announced. They included a commitment by the government to make available £2.5 *billion* to meet any future unforeseen costs of nuclear waste disposal. In many ways this simply represented the continuation of past practice, where government subsidies have protected the consumer from the costs of nuclear waste disposal. Nevertheless, the Energy department's statement was notable for its sheer size, and because of its commendable frankness in making explicit the previously secret extent of the taxpayer's financial commitment to subsidizing the nuclear industry.

This subsidy is justified by the government on the grounds of the strategic necessity of developing the nuclear industry and thereby diversifying sources of electricity supply. Yet, if the strategic argument allows the demolition of the 'polluter pays' principle with regards to the nuclear industry, why does the preservation of Britain's major domestic energy resource not attract similar attention? It would only cost the government a few million pounds a year to preserve the Durham coalfield *and* impose the environmental standards it says it is committed to. The coal industry is denied this relatively small sum, not because the government is committed to an abstract principle, but because of a definite political choice.

Furthermore, in some ways the government seems to have overriden the 'polluter pays' principle, even with regard to the coal industry. For example, the Department of the Environment is spending £5 million over two years funding an experimental project in Walsall to dispose of colliery and power-station waste in flooded limestone mines. The spoil is made into a paste and injected underground (*Mining Magazine*, March 1988). If money is available to subsidize this dumping, why can it not be made available to fund alternative dumping in Durham?

Further contradictions within government policy are not hard to find. Although the government argues that the taxpayer should not pay for alternative dumping options, under present policy the taxpayer will end up footing the bill anyway. When the coastal collieries finally close, the government will help fund the clean-up operation on the beaches through derelict land grants to the local authority. Hundreds of thousands of pounds have already been spent in derelict land grants, reclaiming former pit heaps in Durham. The taxpayer either pays now or later.

Under present government policy the polluter will *never* pay for the damage it has caused and is causing, because it is not commercially possible for British Coal to do so. It would seem to be a depressing and unimaginative government policy which can offer the people of East Durham only two choices: mass job loss or continued environmental destruction.

Alternative dumping sites

Even when presented with possible alternative dumping sites, the government has dragged its heels and misled protestors. When William Thompson suggested the use of Hawthorn quarry as an alternative dumping site, he received this reply from R.A. Hathaway at MAFF, dated 27 March 1987:

> The information available to me is that, far from being disused, this is in fact a working limestone quarry with a long life ahead of it... British Coal advised me that they did in fact consider this quarry as a possible disposal site at one time, but that they were firmly told that this would be totally unacceptable for the reason I have given.

William Thompson wrote back to challenge this assertion, on the basis that the quarry was deserted and there was no machinery left. D.H. Griffiths replied from MAFF on 15 May 1987, with some subtle changes to the Ministry's position:

> The County Planning Officer at Durham County Council ... has advised us that, although it is the case that the quarry is not being worked at the present time, it is in no sense abandoned or disused. The company responsible for working the quarry, who are interested in extracting dolomite for use in glass manufacture, apparently intend to reopen it just as soon as the present slack market for these products picks up again.

However, following letters between MAFF and Durham County Council, another variation in the tale finally emerged. R.A. Hathaway wrote on 3 November 1987:

> We were told that the quarry is no longer in use and that all plant has been removed, although the company holds a lease until the early 1990s. However, the quarry's owners have approached the council to ask if the latter would be interested in using it for tipping, and the council are presently assessing its possible use for domestic refuse.

Either Durham County Council or MAFF misled Seaham residents in these letters.

The significance of Hawthorn quarry is that it lies midway between Easington and Seaham (the two main dumping sites), and only a few yards from British Rail's east coast line. Its use would involve sterilization of dolomite, but at present there appears to be little intention to quarry it. Although the quarry is not large enough to take all the waste from the three coastal collieries over their lifetime (and Durham County Council are vehemently opposed to infilling above former ground levels), it could provide an alternative location, especially in tandem with other more expensive options such as Thrislington quarry.

Other options

Other pits across the country do not dump into the sea. Their waste is taken to quarries, settling ponds and other large holes in the ground. The argument for Durham pits dumping in the sea is that, because they are marginal collieries, any increase in costs threatens their economic survival. Pits elsewhere presumably bear higher dumping costs, but are less marginal and can therefore stand the higher costs. In the absence of any supporting figures from British Coal, it is difficult to challenge this argument.

Of the other options available in Durham, only barging out to sea has been seriously considered. However, MAFF appear very unwilling to license any further offshore dumping following major criticism of their current barging licence for power-station ash from Blyth power-station. As well as barging there are other, more imaginative, options available. An innovative plan has been proposed to use colliery spoil as part of a coastal protection scheme off the Holderness coast. British Coal and the Department of the Environment have agreed to jointly fund research into the viability of the scheme. We have already seen that a scheme is underway to inject waste into underground mine workings. Research into these and other schemes should be expanded immediately. And if the government is prepared to fund research into alternative options, they should also fund the extra cost of inland dumping until research bears fruit.

Durham County Council's favoured option is for dumping to be moved to Thrislington quarry. Both the Ove Arup study carried out in 1985 for the Department of the Environment, and the 1988 consultancy report commissioned by British Coal, identified Thrislington as comfortably the most expensive of the options under consideration (Durham County Council 1989). Yet it is the only option which can accommodate the waste expected over the lifetime of the remaining collieries. The fact is that it is unlikely there is any alternative available to British Coal that is cheaper and more convenient than dumping waste over the cliffs at Dawdon and Easington. Any business forced to cut costs to the absolute minimum will stick to the cheapest waste disposal it can. Without a financial commitment from government to back up ritual acknowledgements of concern, British Coal will continue to use the cheapest, most convenient, dumping option available, which is the North Sea.

SUMMARY OF ARGUMENTS

At present there is a simple dilemma surrounding the destruction of Durham's North Sea beaches and inshore marine environment. To use alternative sites would cost more money. British Coal says that if they had to bear this extra cost it would threaten the economic viability of the dumping collieries. BC's position reflects the narrow definitions of operating viability which the government has imposed on the industry. The government insists that the polluter (and ultimately the consumer) must pay for any alternative; they will not foot the bill.

However, the government's argument is based on a political choice, not on a fundamental principle. The nuclear industry suffers no such financial restrictions, and has indeed been offered £2.5 billion to help dispose of waste products in the future. A further consideration in the government's argument is their deployment of the argument that the taxpayer should not pay for waste disposal. When the mines close, the taxpayer will foot the bill through derelict land grants to the local authorities. Finally, in other parts of the country the government *is* helping fund alternative dumping options.

Although by the end of 1989 British Coal's north-eastern pits were among the most profitable in the country, the economic arguments against BC taking on additional dumping costs remained unrelenting. Electricity privatization will ensure that the Durham pits' economic future remains precarious. In an area of continuing high and chronic unemployment, the threat of further job loss quickly forces most critics to turn their backs on the blighted beaches. With little Conservative interest in this unshakable Labour stronghold, the odds are against a government minister announcing the kind of financial aid that would break the current deadlock. Nevertheless, it is the central conclusion of this chapter that the government *should* pay the extra costs of inland dumping, whether it be to Hawthorn or Thrislington quarries. In doing so they would begin the process of restoring the Durham shoreline to its natural state, and at the same time preserve vital jobs and coal capacity in the Durham coalfield.

The dispute continues

At the end of June 1990 MAFF again renewed British Coal's licence for dumping solid waste over the cliffs. The amount dropped to 1.5 million tonnes per annum, and with the announced closure of Dawdon colliery in 1991, the amount of waste dumped will drop further to around 1 million tonnes. MAFF added a rider to the one-year licence, stating that unless it can be shown that there are 'no practicable alternatives' to cliff dumping, then in five years the licence would not be renewed. British Coal were pressed urgently to seek a land-based alternative. It seems certain that British Coal will seek to rely on the absence of 'practicable alternatives' to continue dumping past the five-year deadline.

The National Rivers Authority have, however, taken a more robust line. The latest 'deemed consent' which they have issued to British Coal is for five years; beyond that period the authority have stipulated that liquid dumping must cease.

Notwithstanding this apparently tough policy, few people expect coastal dumping of coal waste to cease until British Coal have closed down the remaining coastal collieries.

The Parliamentary Ombudsman's report into MAFF's regulation of solid waste licences is still awaited.

The local ombudsman's report into the control of liquid waste disposal concluded that, whilst more lorries than expected were taking liquid waste to the tanker disposal point, they were not necessarily all full, and therefore limits were not being breached.

However, the whole matter has been taken up by Carlos Ripa di Meana and the European Environmental Commission. They are considering whether the dumping is in defiance of Britain's commitments at the North Sea Conference.

REFERENCES

Durham County Council (1983) *Is it a pipedream?*
Durham County Council (1989) *Report of a Joint Officer Working Party.*
Ministry of Agriculture, Fisheries and Food (MAFF) (1979) *Fisheries Technical Report no. 51,* London: MAFF.
McCloskey, G. (1986) *The International Coal Market: a Comparison with British Production Costs and Prices,* Memorandum 82 to the House of Commons Energy Committee Session 1985–6 on the coal industry, relating to the First Report of the Energy Committee 1986–7. House of Commons 165, London: HMSO.
McCloskey, G. and Prior, M. (1988) *Coal on the Market,* London: Financial Times Business Information Ltd.
Mining Magazine (1988) March issue.

Chapter 12

Here today, there tomorrow: the politics of hazardous waste transport and disposal

Denis Smith and Andrew Blowers

A GROWING PROBLEM

> a cargo of hazardous wastes crosses a national frontier more than once every five minutes, 24 hours a day, 365 days per year.
>
> (OECD 1988)

During the 1980s a series of events focused attention on the problems associated with the movements, intentional or otherwise, of hazardous materials across national boundaries. The Chernobyl accident graphically illustrated the potential hazard range of airborne radionuclides and highlighted the difficulties involved in responding to crises of that order of magnitude. The sinking of *The Herald of Free Enterprise* in the English Channel pointed to the potential dangers of transporting hazardous cargoes and the problems involved in taking emergency action following an accident at sea. During June–August 1988 conflicts emerged between Italy and Nigeria concerning the export of hazardous waste from Europe to Africa and this raised questions over the means by which such hazardous transfers are monitored and regulated. The attempts made by the *Karin B* in 1988 to land its toxic cargo in the UK aroused concern about hazardous waste imports, and in the following year cargoes of polychlorinated biphenyls (PCB) waste from Canada were refused entry to British ports. These incidents demonstrated the international character of such transfers and drew particular attention to the routine nature of the trade in wastes. In the context of increasing environmental concerns in the late 1980s, such incidents attracted widespread public and political attention. The aim of this chapter is to review these developments and to consider their political implications.

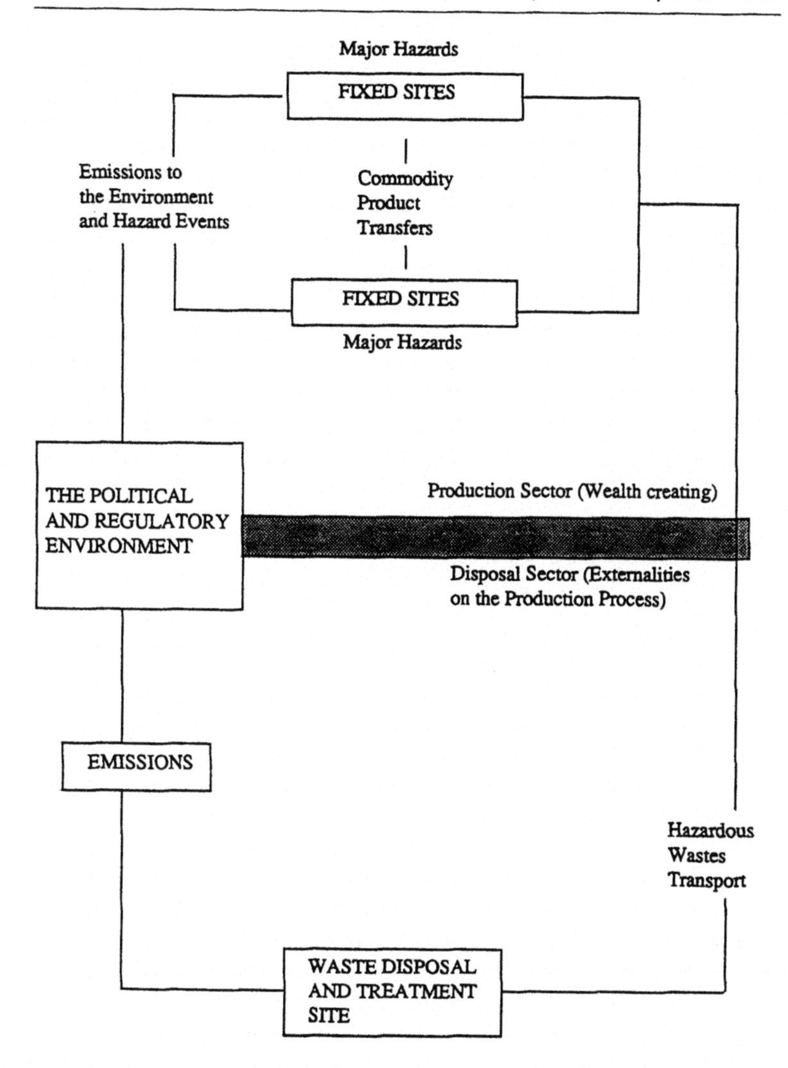

Figure 12.1 Waste transfers
Source: Smith (1990b)

THE NATURE OF THE PROBLEM

The transfer of hazardous materials can be seen as one component of the management of dangerous technologies, forming a link between production sites and the eventual disposal of materials. It is a necessary facet of industrial production that raw materials, finished products and waste residues are transferred between sites as illustrated in Figure 12.1. For the nuclear and chemicals industries these materials are often inherently hazardous and require greater care than would be expected in other industrial sectors. However, the production processes used in these sectors usually create significant local employment and income and thus tend not to attract the degree of local political opposition that might otherwise be expected. Hazardous waste disposal facilities on the other hand tend to attract few jobs and may, in certain cases, discourage new investment in an area. The connotations of 'waste' (which is perceived as having no value) and its attendant hazards has provoked public opposition at many disposal sites. Opposition to nuclear waste disposal sites is widespread and, as in the UK during 1983–7, often successful (Blowers *et al.* 1991). In addition, concern about toxic waste disposal sites, already widespread in the USA and Western Europe, is growing rapidly in the UK.

The issues that concern the public range from amenity damage and economic blight, through the impact on public health, to the long-term damage caused to the local and wider environment by pollution and contamination of soil and water. The problem of hazardous wastes is perceived to extend over both wide geographical areas and time, with the ultimate costs of disposal being borne by future generations. Political opposition always includes localized pressure against specific hazardous waste sites and, in the case of prospective sites, there are usually strong 'Not In My Back Yard' (NIMBY) overtones. Increasingly, there is opposition from nationally-based environmental groups focusing on both the physical damage and health dangers to the wider community and this is combining with local pressure groups to present a formidable challenge to the disposal companies and regulatory authorities. Such combined pressure has not yet achieved the degree of cohesion, expertise and tactical effectiveness demonstrated by the opponents of nuclear waste dumps, although there is evidence of a growing network of organized opposition.

Political reactions to hazardous waste transfer and disposal are partly explained by the nature of the issue. In the first place there is the *ambiguity* of the problem. The hazards from the variety of toxic materials are much less well known than is the case with radioactivity; although public concern has begun to reveal the lack of a universally accepted classification and has highlighted ineffective control of hazardous waste transfers and disposal. Attempts have been made, at national and international levels, to classify wastes according to their properties. A United Nations Committee has detailed nine such categories and the EEC has introduced the Directive on Transboundary Transfers of Hazardous Wastes to overcome the problem of inadequate information and control. Despite these attempts there is still significant ambiguity about the nature of what constitutes hazardous waste, and this problem serves to heighten the political concern over the impact of such waste streams. The lack of adequate information, classification and monitoring also conceals the extent of the problem and restricts and inhibits public opposition. Opposition is also restricted by the difficulties encountered in collecting and analysing data with which to challenge the technical expertise of the waste management companies. The role of regulatory agencies in this respect has been criticized as generally serving a technocratic elite (Fischer 1990; Smith 1990a) rather than promoting more meaningful policy debates between opposing viewpoints.

A second reason for the relative lack of political interest in hazardous waste lies in the relative *invisibility* of the problem by comparison with pollution. A physical distinction can be made between pollution and waste. Pollution is discharged from a point source and dispersed into water or the atmosphere. Waste, on the other hand, whilst still a by-product of the production process, is normally disposed of in a concentrated form, often within some form of containment, or treated to reduce its bulk or toxicity prior to disposal (Wynne 1985). However, the process by which the waste stream is managed and ultimately disposed of may itself generate chronic problems if loss of containment occurs. Further pollution streams can originate from both landfill sites (as pollutants leach into water courses) and from installations which treat waste, such as incinerators (which may release pollutant residues as liquids, solids and via the stack if the incineration process is not totally efficient). More expensive methods of disposal involve neutralization, immobilization and chemical treatment, each of which has the potential to generate pollution streams unless carefully monitored and controlled.

Public opposition may prove difficult to mobilize in those situations where pollution to waste treatment sites has been widely dispersed and the source is difficult to locate. Similarly, the source and culprit may be difficult to trace where pollution occurs long after disposal and the site has subsequently been 'reclaimed'. On the other hand, political hostility is likely to be aroused where there is evidence of neglect or inadequate management during the operational life of the site.

A third factor influencing political attitudes is *profitability*, the paradox that hazardous waste is at once both an unwanted but commercially profitable commodity. The notion of value thus becomes an important concept in the transfer of hazardous materials. Wastes are generally perceived as having no value by the public, although the disposal of waste is a profitable enterprise. The costs of waste disposal are often discounted by industry in the early stages of product development in order to improve the economic effectiveness of a particular proposal (see, for example, Elkington and Burke 1989). The implication here is that unless the polluter is forced to pay for the use of the environment as a pollution sink, then the cost of degradation will not show as a significant cost within the production process (see Pearce *et al.* 1989 for a full account of this problem).

Companies will usually seek the cheapest means of disposal and this can be by country or by companies which offer facilities at competitive rates. Costs of disposing municipal wastes are indicative of the differences that can exist. In 1988 it was estimated that waste disposal in New York would cost £70 per tonne; in Philadelphia £26.50; but in England and Wales only £5 (Hewitt 1988). Such differences encourage trade, and waste therefore becomes a commercial commodity. As a consequence of the considerable profits which can be obtained through waste disposal, loopholes have often been found in the legislation to overcome potential barriers. In West Germany, for example, a considerable amount of hazardous waste has been exported as a valued product, termed 'Wirtschaftsgut', thereby avoiding much of the stringent legislation reserved for waste products in that country (Schurmans 1988).

Exporting hazardous wastes to countries outside the affluent economic groupings may be both cheap and politically expedient. Developing countries have poorly-developed technical expertise, an absence of effective regulatory mechanisms, and inadequate accounting procedures, thus allowing for the relative ease of illegal dumping. The problem of monitoring waste is, however, not restricted to the developing nations. In Japan, for example, it has been suggested that the official estimates of hazardous waste totals may well be at least a factor often too low (Ui 1984). In the UK the corporate origins of imported wastes were often unknown prior to 1989 (Department of the Environment 1988) and this raised issues of liability in the event of a major polluting incident and undermined public confidence in the regulatory process.

Recently the transfer and disposal of hazardous wastes has become less ambiguous, less invisible and, in some respects, less profitable. In a period of heightened environmental awareness, communities and countries are demanding better surveillance of hazards, more information and insisting on companies taking responsibility from the cradle to the grave. The transfer of hazardous wastes has become highly politicized and the United States can be used to illustrate this process of politicization.

THE POLITICIZATION OF TOXIC WASTE IN THE UNITED STATES

Estimates suggest that the USA generates some 266 million tonnes of hazardous waste a year of which only 10 per cent is considered to be disposed of in a manner which is deemed to be environmentally acceptable (see, for example, Wright 1986; Newsday 1989). Over the period 1973–80, hazardous waste emerged as a major political issue in the USA with public opinion indicating widespread objections to having hazardous waste dumps located close to residential areas (Porcade 1984). The dramatic shift in public awareness and concern that occurred during this period can be accounted for by the problems experienced at Love Canal, close to Niagara Falls, in the late 1970s.

When the Hooker Chemical Company completed the disposal of 21,000 tonnes of toxic wastes at the site during the 1940s the dump was 'sealed' and 'reclaimed' and sold to the local school board. The board later sold parcels of land for building development, although this was contrary to the agreement with Hooker Chemical. The subsequent development of a new residential area, completed in the 1970s, breached the membrane of the disposal site and eventually led to leaching of toxic materials which provided an immediately hostile environment for local residents (Harthill 1984). One account describes,

> patches where the chemicals were concentrated enough to prevent vegetation from growing ... On the surface, one could see patches of coloured chemicals. In some places, chunks of solid chemical waste surfaced ... In several spots, holes appeared

which filled with black oily leachate, presumably where 55 gallon drums under the surface have disintegrated.

(Paigen 1984: 3)

A number of pollution incidents occurred and the local community were fearful of the health effects presumed to be caused by the seepage of toxic fumes and leachate into their homes. There was some evidence of an increase in foetal deaths, chromosone damage and a cancer rate thirty times the national average (Brown 1980). Whilst causality was difficult to prove, the fear expressed by local residents ensured that Love Canal became a notorious political issue in the late 1970s. Eventually the whole area was evacuated and Love Canal remains today boarded up and silent with warning notices of the lurking danger beside the fenced-off grassed-over area of the former disposal site.

The initial impact of Love Canal was to undermine the conventional wisdom that such sites were reclaimable. The episode illustrated that it was simply not enough to bury the waste in the hope that once it was 'out of sight' it would be 'out of mind'. The longevity of the chemical substances buried at the dump site meant that the questions of liability and accountability over hazardous waste were

Table 12.1 Hazardous waste conflicts in the USA (selected events)

Date	*Place*	*Event*
1980	Swartz Creek	12,000 tonnes of contaminated waste poorly disposed
1981	Stringfellow Acid Pits	34 million gallons of waste stored in poor conditions
1983	Times Beach, Missouri	Town evacuated due to dioxin contamination (bought by EPA)
1983	Rocky Mountain Arsenal	Costs in excess of US $4 billion for clean-up
1984	Chem. Waste Wells, Ohio	Leak of toxics into watercourse
1985	Bridgeport, New Jersey	Waste site reclamation to US $50–60 million
up to 1984		EPA designated 786 waste sites under SUPERFUND – total projected to rise to 2,500

Source: Boraiko, 1985.

not easily determined. As a consequence, federal involvement was requested to deal with the problem which brought about a commensurate increase in political concern, culminating in the setting up of a fund by Congress in 1980 (known as the 'Superfund') to clean up such sites. Although the intentions of this programme were well founded, the scale of the problem in the USA has proved to be too great to deal with in the short term, with up to 10,000 sites being in need of treatment. Given that from 1980 to 1986 the clean-up rate has averaged only one site a year some writers have expressed grave concern over the future effectiveness of such measures (Wright 1986).

The costs involved in the clean-up of such sites are substantial, as the case of the Rocky Mountain Arsenal site illustrates. The Rocky Mountain Arsenal consists of a 27-square-mile site in which the US Army and Shell disposed of hazardous waste (mainly pesticides, in the case of Shell and the Army's chemical warfare waste). The site operated from 1950–82, although Shell only became involved in 1962, and the estimated clean-up costs are thought to be between US $2–4 billion and will involve extensive on-site treatment and incineration of wastes. Of these costs, approximately US $1 billion will be borne by Shell with the remaining liability being met by the US Army (Nicod 1990).

Whilst Love Canal raised public awareness of the hazardous waste issue, a number of other events over the period 1978–84 served to keep the issue high on the political agenda in the USA (see Table 12.1). The mounting concern led to the resignation of Environmental Protection Agency (EPA) staff in 1983 over the delay in using the Superfund to clear problem sites (Boraiko 1985). The outcome of these events was that it became increasingly difficult to dispose of all forms of waste in the USA owing to local public opposition (Craig and Lash 1984). Similar opposition also emerged in a number of European countries and, in certain circumstances, has prompted a move to consider the transfer of hazardous waste to other countries. Thus the politics of hazardous waste transfer has local, national and international dimensions, and these need to be assessed in the light of recent events.

THE POLITICAL ENVIRONMENT OF HAZARDOUS TRANSFERS

There are four reasons for transfrontier shipments of hazardous wastes (Mills 1988). First is the inadequate treatment or disposal facilities at the point of waste arising. For instance, some countries, like Holland, lack the space or the geology for adequate landfill disposal. Second, the nearest treatment facilities may be across a frontier. Third, as we have indicated, the costs of treatment vary and cheap facilities may justify the cost of long-distance shipment. Finally, regulatory controls also vary and companies will naturally seek out those countries where controls are most relaxed or widely evaded if this brings down costs. Economic and regulatory criteria are likely therefore to be most important in determining shipments between developed and developing countries. More recently political opposition at the point of waste arising or disposal, forcing companies to search for alternative disposal sites outside the region or country of origin, is becoming a major factor in the geography of waste transfers.

The political conflict over hazardous wastes takes place between different sets of interests, both economic and environmental, and at different levels, local and national. Typically, within the developed countries, local communities mobilize opposition against companies and sometimes the central government. Corporate interests are able to exert power through their investment in the community, through the jobs they create or through the promise of prosperity. They may even succeed simply through inaction, although such a course is less likely to be effective now that the issue has become politicized. If their local economic influence is insufficient to influence policy decisions, then they can appeal to central government in an attempt to secure a favourable outcome. The various corporate strategies of power in dealing with environmental issues have been extensively discussed in the literature (for references on polluting industries see Crenson 1971, and Blowers 1984, and for major hazards see Smith 1990a).

As costs have increased, regulations tightened and the political climate has become more hostile, companies have found themselves squeezed out of disposal sites in the developed countries. Increasingly they have turned their attention to the cheaper, more tolerant prospects in the developing world. It is the lower costs of disposal and lack of regulatory frameworks that exist in developing countries which have encouraged a number of western companies to export waste for landfill. The companies involved in such movements may be of two types. The first group consists of the large chemical transnational corporations (TNCs) which may have a subsidiary in the country where the waste is to be disposed. In this case the waste can either be that which arises as a result of the activities carried out within the country; or, alternatively, can have arisen from another division of the TNC operating elsewhere. The second group of companies are those which have developed as specialized hazardous waste firms and who operate through a network of waste brokers. These companies take waste from larger corporations and then seek out low-cost repositories for it. Because of their size, these companies are often politically unaccountable, as was illustrated during 1988 by the dumping of Italian hazardous waste in West Africa.

It is possible to identify a hierarchy of national types developing between states involved in the transfer of wastes. Figure 12.2 shows this relationship in terms of a '*pyramid of exploitation*' (Smith 1990b). Within this pyramid it is possible to categorize countries in terms of the level of their industrial development. The resulting categories are shown in Figure 12.2 and can be considered as core, semi-core and peripheral nations. As countries move upwards through the pyramid they achieve increased technological development and an increasingly stringent regulatory framework, both of which are fundamental in moving a nation away from using such low-tech methods of waste disposal as landfill. Growing local political opposition to landfill and incineration sites will prompt corporate groups either to export wastes or, alternatively, to adopt waste minimization technologies. Those nations which are found in the top two categories are likely to provide a base for TNCs which transcend national boundaries and are a prime factor in the global spread of capital. It could be argued that TNCs are an intrinsic factor in what can be termed the '*cascading exploitation*' of the peripheral nations (including developing countries and those industrial nations which are economically stagnant). The ability of TNCs to circumvent legislation by moving their operations to another country enables them to exploit those countries which are desperate for foreign capital (Ives 1985). In the most developed nations, on the other hand, more sophisticated technology provides a waste management service that is not universally available (for example, immobilization or high-temperature incineration). There is, therefore, a trade in hazardous wastes between industrialized countries, and also an export trade from developed to developing nations both of which require regulating in order to prevent major environmental problems.

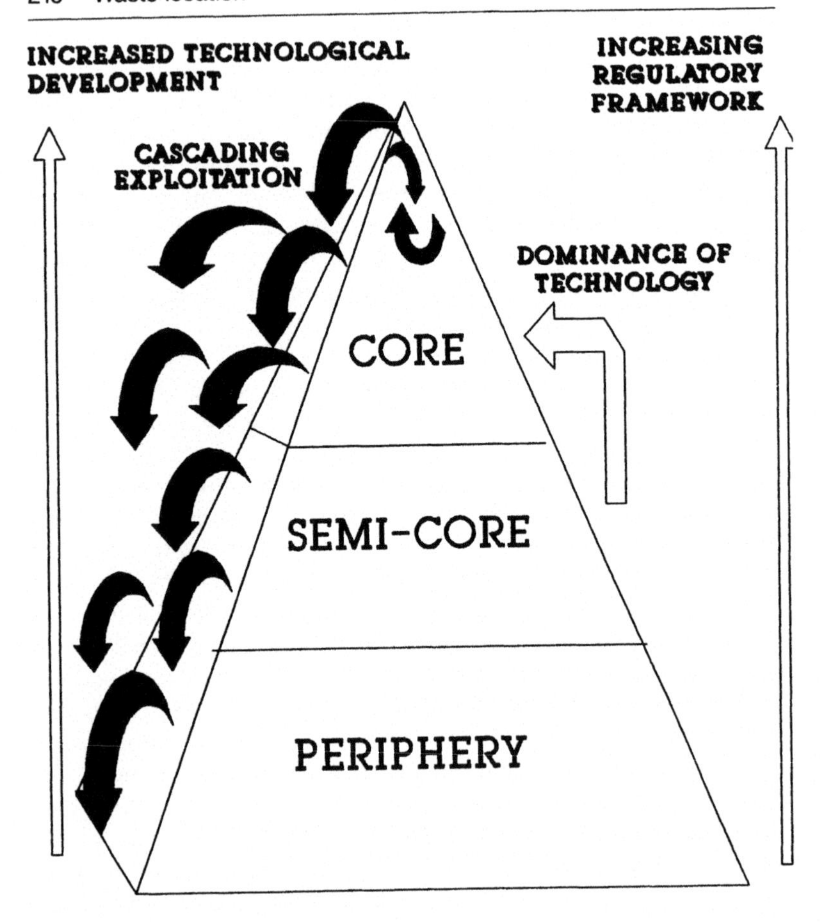

Figure 12.2 The pyramid of exploitation
Source: Smith (1990b)

THE TRANSFER OF HAZARDOUS WASTES IN EUROPE

The transfer of hazardous wastes in Europe has become a significant political issue as governments become increasingly reluctant to accept waste from other countries. Until 1988 there was a lack of a common definition of hazardous wastes within the EC and this, combined with a lack of effective monitoring, produced a confused picture of waste transfers. Effective regulation has proved difficult both to introduce and to enforce. Despite recent EC directives there are still problems of definition, notification and monitoring of shipments in European countries. The problems stem from the size of the market in Europe, the generally under-resourced nature of the regulatory agencies and the multinational character of the chemicals industry which is responsible for most of the transfers. It is estimated that 100,000 waste transfers occur within the EC annually (Laurence and Wynne 1989). As the Community moves towards an integrated market after 1992, greater efforts will be required to control such movements. A brief examination of the situation in the Netherlands and the UK illustrates some of the problems.

The Netherlands is particularly affected by European transfers since it receives large quantities of hazardous waste across its borders, notably from (West) Germany, and then exports some of it to the UK. In Holland, the presence of a high water-table, the absence of underlying clay strata, and the low-lying nature of the topography render new landfill proposals largely impracticable except at high cost. As a consequence, the Netherlands has been forced to develop its technology for high-temperature incineration and immobilization and only aims to export waste (mainly to the UK, although some was sent to what was East Germany) for landfill. However, in order to avoid the political sensitivity of being totally dependent on other countries, the Dutch Government has developed a landfill site at Rijnmond, close to Rotterdam. The ultimate aim of this strategy is to make the Netherlands completely self-sufficient in terms of waste disposal capacity.

In the UK landfill is still by far the most common disposal route with around 85 per cent of waste being dealt with in this way. Many such sites would not be considered satisfactory in the United States nor within some European countries, notably (West) Germany, the Netherlands and France (Hildyard 1989). Set within this context, the small number of inspectors within the UK (six to monitor some 5,000 sites in 1990) has been a cause for concern as has the apparent reluctance of some waste disposal authorities (WDAs) to implement the requirement to produce detailed plans for the disposal of hazardous waste (House of Commons 1989).

As landfill capacity becomes scarce and expensive in North America and Western Europe, the UK has become an attractive haven for hazardous wastes. Landfill is the preferred technique in the UK where there are large numbers of suitable clay-lined sites, remnants of industries such as brick-making, which afford site operators with the opportunity to make significant profits from their investment in the land. Waste disposal companies have invested heavily in these sites and attract waste from far afield, both within the UK and from the continent. As an indication of the scale of the problem, imports of special wastes to the UK rose from 3,800 tonnes in 1981 to around 80,000 tonnes in 1988 but declined to 40,000 tonnes in 1989. This decline seems likely to continue as the government's proposed environmental legislation seeks to curtail such trade for direct landfill, although imports for other forms of treatment will not be affected.

Politically the greatest concern in the UK has centred on the high-temperature chemical incineration processes which deal with the more persistent and hazardous substances such as PCBs. The country had a capacity in 1991 of some 73,000 tonnes per year in three incinerators. As waste management regulations tighten, more wastes will have to be treated by incineration. The trade in hazardous wastes for treatment seems certain to continue and will bring with it continued political tensions. One such example was the incident in the summer of 1989 when attempts to import Canadian PCB waste for disposal by high-temperature incineration aroused considerable public opposition. The waste was the result of a fire in a warehouse in Quebec and it was shipped to the UK aboard three Soviet vessels between July and August. The first vessel, the *Kudozhnik Saryam*, encountered serious opposition at Tilbury and was refused permission to unload. Similar opposition at Liverpool eventually ensured that all three ships returned with their cargoes to Canada. The cargoes were intended for the incineration facility operated by ReChem at Pontypool in South Wales. Although the vessels were prevented from discharging their cargoes by the actions of dock workers and port authorities, the trade was lawful. By refusing entry of hazardous wastes for effective disposal in the UK, protesters may eventually ensure such materials are indiscriminately dumped in a Third World country.

THE EXPORT OF WASTE TO THE DEVELOPING WORLD

The extent of proposed and existing hazardous waste transfers to the developing countries during the late 1980s is indicated in Table 12.2. Whilst most of this waste came from industrial and chemical processes, France and the USA also attempted to dump radioactive waste and uranium mining tailings respectively. Guinea-Bissau has been a major destination for wastes from developed countries and was offered the equivalent of its GNP (some US $120 m) to dispose of European hazardous waste in landfills. Such offers are difficult to

Table 12.2 Proposed exports of waste to West Africa (1986–9)

Country	*Waste quantity and type*	*Country of origin*	*Value*
Nigeria	3,800 tonnes	Italy	$250 per month
Guinea	15,000 tonnes incinerator ash	USA/Norway	$600,000
Benin	>5 million tonnes	USA/Europe	$12.5 million
Equatorial Guinea	2 million tonnes	Europe	?
Gabon	Uranium mining waste	USA	?
Congo	1 million tonnes solvents and chemicals	USA/Europe June 1988–May 1989	$4 million
Guinea-Bissau	15 million tonnes chemicals	UK/USA/ Switzerland terminated in June 1988	$120 million per year
	15 million tonnes chemicals	Contract cancelled	

Sources: Secrett (1988), Wynne (1989) and Third World Network (1989)

refuse given that the economic benefits are immediate whilst the physical costs may not be borne for many years. While most of the political attention has been focused on western exports, there have also been transfers of wastes between developing countries (Third World Network 1989). European governments have agreed to severely restrict further exports to the developing world but the inadequate controls, inadequate regulatory frameworks and low costs remain powerful incentives to maintain the trade. 'Huge profits are at stake, with governments and others in these countries being offered a few dollars per tonne to accept wastes which would cost up to £1000 per tonne to dispose of in the industrialised world' (ENDS 1988: 9). Indeed, a network of waste 'brokers' is already operating in Europe and these economic incentives will be prime factors in encouraging such entrepreneurs towards profit maximization by seeking a Third World location for waste.

It was the voyage of the *Karin B* during 1988 which drew international attention to this trade in hazardous wastes. About 4,000 tonnes of mixed chemical waste, including PCBs and some radioactive material, had been exported from Italy and stored on a site in Koko, Nigeria, at the end of 1987 (Third World Network 1989). The waste was held in over 2,000 drums, some of which were leaking, and included 150 tonnes of highly-toxic PCBs. The local landowner was paid a mere US $250 per month to store the material and was generally unaware of the risks that he was taking (Third World Network 1988). The revelation of the existence of this dump was embarrassing for Nigeria which had earlier been critical of a decision to dump waste in neighbouring Guinea and Guinea-Bissau and had inspired the Organisation for African Unity to pass a resolution condemning the disposal of hazardous wastes in Africa (Harden 1988). Nigeria's response, in the wake of a potential political backlash, was swift and forceful. In June 1988 it was announced that the town of Koko was to be evacuated and that the importers, if caught, would be shot (Reuter 1988). Nigeria recalled its ambassador in Italy and also confined an Italian vessel at anchor in Lagos (Hiltzik 1988). Following an intense period of publicity in the international press, the waste was removed from Koko aboard the *Karin B* and the *Deep Sea Carrier* and the two ships began their long voyages in an attempt to discharge their cargo. Refused entry to France, the UK, West Germany and Spain, the *Karin B* returned to Italy where the cargo was eventually unloaded at Ravenna in December 1988 (Wynne 1989).

Two other vessels at sea during July–August 1988 also found difficulty in landing their cargoes, the *Zanoobia*, also carrying Italian waste, and the *Khian Sea*, the so-called 'leper of the oceans'. The *Khian Sea* eventually disposed of its cargo in mysterious circumstances and in November 1988 arrived off Singapore, unladen, with a new name (*Pelicano*) and in new ownership (Wynne 1989). In September 1990 it was announced that waste from the *Zanoobia* was being processed at Cadishead in Manchester (Raphael 1990). Like the *Khian Sea*, the wanderings of the *Zanoobia* reflect the complex nature of such transfers. The waste originated in Italy and was initially exported to Djibouti although it was subsequently routed to Venezuela after the port authorities in Djibouti refused it entry. Following opposition to the waste in Venezuela it was then exported to Syria and ultimately returned to Italy (Raphael 1990). From there part of the cargo was exported to the UK for treatment and disposal. The wanderings of these various ships carrying unwanted cargoes gave world-wide attention to the problem of international trade in hazardous wastes and ensured that the issue would remain high on the political agenda into the 1990s.

CONCLUSIONS

During the last decade hazardous wastes have become an issue of international political concern. The politicization began in the United States where incidents such as Love Canal stirred public indignation spread to western Europe with its complex system of internal transfers and finally embraced the developing world, notably Africa, which was becoming a dumping ground for toxic substances from the wealthier nations. The pattern of transfrontier transfer reflects fundamentally technical, economic and political inequalities. In the best-regulated industrialized countries waste streams are separated, landfills carefully managed and hazardous wastes are incinerated or immobilized whenever possible. In the developing world mixed wastes tend to be dumped indiscriminately with little regard to environmental or health impacts. The high cost of disposal in industrial countries, combined with the willingness of poorer countries to accept wastes for the currency they will bring, has encouraged long-distance transfer. Whereas political concern in the western nations has led to tighter regulations, elsewhere the environment and public health is of lesser significance when economic survival is at stake.

There are signs, however, that the picture is changing. Growing environmental concern within western countries is focusing more on global problems and this, coupled with the reaction of Third World countries against exploitation and environmental degradation, is likely to outlaw much of the trade. As it does so, the options for waste management will reduce and solutions to the problem will have to be sought at an international level. Such solutions will have to focus on the three components of the problem we identified at the outset. First, *ambiguity* about the nature of waste will have to be reduced. A concerted international effort will have to be mounted to achieve agreed technical criteria for waste management. This will need to combine with an effective monitoring system and enforcement and sanctions procedure which is sufficient to deter cowboy operators and irresponsible countries. Second, greater international control mechanisms can only be effective if the *visibility* of the problem of hazardous wastes is increased. The sources of wastes, their environmental impacts and conditions of treatment and disposal must be identified so that responsibilities for sites can be ensured. Third, *profitability* must incorporate appropriate funding for environmental safety and also provide for the security of disposal extending down the generations. If the 'polluter pays' principle is strictly adhered to (within a strong legislative framework), and full costs are imposed upon companies, then the exploitation of the poorer countries will be severely curtailed. If these various conditions are brought into operation then the transfer of hazardous wastes will reflect the availability of properly managed, politically accountable and economic sites rather than weaknesses within the system. Hazardous waste will then become a commodity that is transferred for reasons of safety and environmental protection rather than for profit.

REFERENCES

Blowers, A.T. (1984) *Something in the Air: Corporate Power and the Environment*, London: Harper & Row.

Blowers, A.T., Lowry, D. and Solomon, B.D. (1991) *The International Politics of Nuclear Waste*, London: Macmillan.

Boraiko, A.A. (1985) 'Storing up trouble ... hazardous waste', *National Geographic* 167 (3), 318–51.

Brown, M. (1980) *Laying Waste: The Poisoning of America by Toxic Chemicals*, New York: Pantheon Books.

Craig, R.W. and Lash, T.R. (1984) 'Siting of radioactive hazardous waste facilities' in M. Harthill (ed.) *Hazardous Waste Management: In Whose Backyard?*, Boulder, Col.: Westview Press, 99–110.

Crenson, M.A. (1971) *The Unpolitics of Pollution: Decision-making in the Cities*, Baltimore, Md: Johns Hopkins University Press.

Department of the Environment (1988) Private communication.

Elkington, J. and Burke, T. (1989) *The Green Capitalists*, London: Victor Gollancz.

ENDS (1988) *Environmental Data Services Ltd*, Report, June.

Fischer, F. (1990) *Technocracy and the Politics of Expertise*, Newbury Park: Sage.

Harden, B. (1988) 'Africa battles to turn back a tide of toxic waste', *International Herald Tribune*, 23 June 1988.

Harthill, M. (1984) 'Introduction', in M. Harthill (ed.) *Hazardous Waste Management: In Whose Backyard?* Boulder, Col.: Westview Press, 1–6.

Hewitt, M.R. (1988) 'The transfrontier directive and the principles of importation of waste within the UK', paper presented to the IWM symposium on *The Importation of Waste to the UK*, Institute of Wastes Management, London, 8 September 1988.

Hildyard, N. (1989) 'Toxic wastes: playing in a fools paradise?', *The Ecologist* 19 (4), 126–8.

Hiltzik, M. (1988) 'Waste scandals add to strain with West', *The Star* 21 June 1988.

House of Commons (1989) Environment Committee, Second Report, Session 1988–89, *Toxic Waste*, London: HMSO.

Ives, J. (ed.) (1985) *The Export of Hazard: Transnational Corporations and Environmental Issues*, Boston: Routledge and Kegan Paul.

Laurence, D. and Wynne, B. (1989) 'The international trade waste in Europe: towards a free market? *Environment* 31 (6), 12–17 and 34–5.

Mills, D.A. (1988) 'The EC directive on the transfrontier shipment of hazardous waste', paper presented to the IWM symposium on *The Importation of Waste to the UK*, Institute of Wastes Management, London, 8 September 1988.

Newsday (1989) *Rush to Burn – Solving America's Garbage Crisis*, Washington, DC: Island Press.

Nicod, J. (1990) 'Liquid interim response action', *Etudes Et Dossiers*, Background paper, Basin F. 146, 147–67.

OECD (Organization for Economic Co-operation and Development) (1988) *Transfrontier Movements of Hazardous Wastes*, Paris: OECD.

Paigen, B.J. (1984) 'Methods of assessing health', in M. Harthill (ed.) *Hazardous Waste Management: In Whose Backyard?* Boulder, Col.: Westview Press, 37–62.

Pearce, D., Markandya, A. and Barbier, E. (1989) *Blueprint for a Green Economy*, London: Earthscan Publications.

Porcade, B.S. (1984) 'Public participation in Siting', in M. Harthill, (ed.), *Hazardous Waste Management: In Whose Backyard?* Boulder, Col.: Westview Press, 111–12.

Raphael, A. (1990) 'Britain supine on murky trade in toxic wastes', *Observer*, 16 September 1990, 9.

Reuter (1988) 'Nigeria gets ready to evacuate radioactive waste town', *New Straits Times*, 15 June 1988.

Schurmans, Y. (1988) 'European trends in hazardous waste disposal or Belgium at the crossroads', paper presented to the IWM symposium on *The Importation of Waste to the UK*, Institute of Wastes Management, London, 8 September 1988.

Secrett, C. (1988) 'Deadly offer poor countries find hard to refuse', the *Guardian*, Friday July 15, 11.

Smith, D. (1990a) 'Corporate power and the politics of uncertainty: conflicts surrounding major hazard plants at Canvey Island', *Industrial Crisis Quarterly* 4, 1–26.

Smith, D. (1990b) 'The international trade in hazardous waste: a study in the geo-politics of risk', paper presented at the European Consortium for Political Research (ECPR) workshop on *Crisis Management*, Bochum, West Germany, 2–7 April 1990.

Third World Network (1988) *Report on Toxic Dumping in Third World Countries*, Penang, Malaysia: Third World Network.

Third World Network (1989) 'Toxic waste dumping in the Third World', *Race and Class* 30 (3), 47–56.

Ui, J. (1984) 'An overview on solid waste and hazardous waste in Japan', *Conservation and Recycling* 7 (2–4), 67–71.

Wright, J.W. (1986) 'Initiating high risk projects in bureaucracies', *Project Appraisal* 1 (3), 160–8.

Wynne, B. (1985) 'A case study: hazardous waste in the European Community', in H. Otway and M. Peltu (1985) (eds) *Regulating Industrial Risks: Science, Hazards and Public Protection*, London: Butterworths, 149–75.

Wynne, B. (1989) 'The toxic waste trade: international regulatory issues and options', *Third World Quarterly* 11 (3), 120–46.

Chapter 13

Narrowing the options: the political geography of waste disposal

Andrew Blowers

GROWTH OF ENVIRONMENTAL CONCERN

During the 1980s the impact of waste disposal on public health and the environment became an issue of increasing political salience. Public alarm grew about the dangers of hazardous waste transfers, the management of radioactive wastes, the pollution of groundwater by toxic chemicals and the discharge of wastes into the sea. The management of wastes and the reduction of pollution were key issues in the growing interest in the environment. A survey conducted in the UK in 1989 revealed that the environment and pollution was regarded by 30 per cent (unprompted) of a sample as an important issue for government to deal with (against only 8 per cent in 1986), second only to health and social services which was mentioned by 32 per cent.[1] Similarly, in the United States, a poll in 1989 indicated that the environment ranked third after education and homelessness as the most pressing problem.[2]

This public concern has been echoed in political statements. The Prime Minister Margaret Thatcher signalled the government's concern about atmospheric pollution in her speech to the Royal Society in 1988.[3] Later, she sought to appropriate the environmental issue for the Conservative Party and in a speech to the 1988 Conservative Party Conference she declaimed, 'No generation has a freehold on this earth. All we have is a life tenancy – with a full repairing lease. And this Government intends to meet the terms of that lease in full.' The rise of the Green Party, taking 15 per cent of the UK votes in the European elections of 1989, and the prominence given to environmental matters by the opposition parties, were spurs to action by the government. The apparent greening of the government has been evident both at the rhetorical level and in the discussion of environmental pricing policies prompted by the Pearce Report (Pearce *et al.* 1989). At a more practical level an Environmental Protection Bill was introduced at the end of 1989 and a White Paper on the environment in 1990. But inaction tended to be greater than words as the government procrastinated over the implementation of sulphur oxide reductions and over the dumping of untreated sewage in the North Sea. In terms of priorities there was little evidence that transportation, energy and industrial policy were being much affected by environmental concerns.

Contemporary concern for the environment is episodic but not ephemeral. The prominence given to environmental issues is likely to subside when more short term economic concerns press upon people and government. But, it is unlikely that environmental concern will disappear. The environment has established a permanent and prominent place on the political agenda. People are able to make the connections between nuclear energy and leukaemia clusters, and they fear the hazards of groundwater pollution, dirty beaches or polluted oceans. There is a growing recognition that contemporary forms of economic and technological development will result in environmental deterioration and degradation in the longer term. The necessity for sustainable development is widely proclaimed, to ensure, in the words of the Bruntland Report, that humanity 'meets the needs of the present without compromising the ability of future generations to meet their own needs' (World Commission on Environment and Development, 1987:8).

THE SOCIAL IMPACT OF ENVIRONMENTAL CHANGE

This book has focused on the location of processes responsible for the production of waste. Wastes are the unwanted or unused residues of organic and industrial processes, which, if they are not carefully managed, cause environmental risk and damage in the form of pollution. Therefore the production and location of waste is inextricably bound up with the distribution and impact of pollution. There are three features of the distribution of waste and pollution which have fundamental political implications. The first is the increase in the *scale* of the problem. Local environmental concerns are, of course, always present, but to these have been added problems such as acid rain that cross political frontiers, those such as depletion of the ozone layer or the increase in atmospheric carbon dioxide which are of global dimensions, and those like radioactive wastes which extend down the future generations.

With this increase in scale has come a second feature, a focus on different *environmental problems* created by waste and pollution. The traditional local concern with the quality of the local environment, the need to protect amenity and public health remain, though they have changed over time. A century ago there was the problem of epidemic diseases spreading among the poor living in the insanitary conditions of the big cities. The squalid conditions had to be improved if only to prevent contagion spreading into the territory of the wealthy in the suburbs and beyond. Today there is the fear of toxic poisons or of radionuclides lurking in the accessible environment and inflicting illness on unsuspecting populations. To these localized concerns has been added a more general concern about the degradation of the environment, the depletion of resources and the survival of habitats, species and ecosystems, indeed, in the long term, a fear for the survival of life on the planet.

The third feature is the changing *social impact* of pollution and waste. In the past environmental pollution tended to be socially selective in its impact. Air and water pollution tended to be localized and could be avoided by those with resources to live elsewhere. The poor bore the brunt of the pollution from the industries on which their livelihoods depended. Environmental conditions reflected and reinforced the patterns of social inequality. The pattern of environmental inequality persists within the developed countries, though the close spatial relationship between work, residence and pollution has been loosened with greater mobility and changing work patterns. The association of social inequality, environmental quality and spatial location also occurs between countries as polluting activities are exported from rich to poor.

As well as the local impacts which result in environmental inequality, there are waste and pollution processes which are less discriminating in their social impact. Economic growth has brought greater mobility enabling mass travel to places once the preserve of the wealthy. Thus attractive environments have become more accessible but in the process more threatened with pollution. Technological developments have increased the scale of pollution and the risks from toxic and radioactive materials affect rich and poor alike. Indeed, on the world scale, the impact of atmospheric pollution threatens a deterioration in the environment from which none can escape. Contemporary environmental concerns are truly both local and global.

THE CHALLENGE FOR GEOGRAPHY

Geography is one of the disciplines concerned with the identification and explanation of environmental change and its consequences. Geographical analysis can also be applied to environmental planning and management. Yet, the contribution of geography to the development of environmental studies has been relatively modest and muted. This is partly because the environment requires interdisciplinary study and an integrated approach to management. Geography's contribution is in co-operation with other disciplines rather than providing specialist expertise. Geography offers a perspective that emphasizes the spatial context within which social and physical environmental processes interact.

There are three elements that make up the geographical perspective. The first is the *social/spatial* element, the concern with spatial relationships. This has developed from an early obsession with functional relationships and statistical methodology to a sophisticated theoretical analysis of the spatial division of labour and empirical locational analysis of such features as industry, transportation and social areas. Locational analysis has become a fruitful field for the application of Geographical Information Systems. Carver and Openshaw (Chapter 7) present an example of the application of the technique to the problem of nuclear repository site selection. The claim that the technique provides an objective basis for political choice is critically discussed by Sheldon and Smith in Chapter 10.

Geography has also been traditionally interested in the interaction between society and the physical environment. This is the *social/natural* element of the geographical perspective. The early debates between the 'possibilists' and 'determinists' centred on the extent to which nature constrained human activities. The idea of environmental determinism is currently resurgent. The social/natural element in geography justified the coexistence of the physical and human branches of the discipline, though they have become increasingly divorced. Geography is not the only discipline focusing on the interaction of people and their physical surroundings. Many other disciplines, notably, medicine, biology and ecology focus on the impact of environment on population, and others such as urban studies, social psychology, architecture and town planning are concerned with the behavioural consequences of people's physical surroundings. An example of the social/natural focus in geographical studies is Gatrell and Lovett's Chapter 9 demonstrating the possible links between the incineration of toxic materials and certain forms of cancer and Raybould *et al.*'s study in Chapter 8 inferring a link between heavy metals in soil and diabetes. Both these chapters illustrate how the social/spatial and social/natural elements can be combined in the effort to identify significant relationships between human and physical phenomena.

These two elements of the geographical perspective may also be combined within a third, the emphasis on *uniqueness and interdependence*. Regional geography has traditionally interpreted the combination of physical and human features which compose the unique characteristics of regions. Systematic approaches have isolated specific processes such as industry, agriculture or resources which link regions together in interdependent systems. In terms of the environment the unique qualities of regions are being threatened by their dependence on the resources of air, land and water. Interdependence results in a shared vulnerability to environmental degradation and pollution. The need to protect unique qualities by careful environmental management is a theme that appeals to geographers. In Chapter 2 Kivell assesses the advantages of using derelict land for waste management. In Chapter 3 Bridges underlines the need to find cover materials which do not destroy the landscape, and in Chapter 11 Renouf considers the economic and political factors responsible for the degradation of the Durham coastline by sea dumping of coal.

These three elements in the geographical perspective provide a focus on locational opportunities and constraints and identify the relative advantages of specific locations. For instance, Carver and Openshaw (Chapter 7) plot geological details, population data, accessibility data and conservation areas in producing the variables that can be mapped in a GIS for locating a nuclear waste repository. Crichton (Chapter 5) used a data base incorporating details of waste-site licences in order to demonstrate the advantages of GIS in matching site characteristics to waste streams. Geographical analysis can also explore the need for data in making appropriate locational choices. For instance, Parfitt (Chapter 4) illustrates the need for adequate information in the selection of waste disposal routes. The geographical concern with interrelationships emphasizes the need to identify and monitor the consequences of decisions. Chapter 6 by Coggins *et al.* shows how the introduction of new waste collection methods imposed unforeseen costs on waste disposal.

In general, geographers rely on rational explanation using

functional data to identify locational opportunity and constraint. This procedure cannot wholly explain specific locational decisions. A vital but neglected element in the geographical perspective is the relationship between *politics* and *place.* The political context of locational decision-making is the major theme of the chapters by Sheldon and Smith (Chapter 10) and Smith and Blowers (Chapter 12). The rest of this chapter outlines why politics must be incorporated into geographical thinking as an essential element in the explanation of locational processes.

POLITICAL CONFLICT AND THE ENVIRONMENT

Politics is about the exercise of power in making decisions involving choices between alternatives. Choice involves conflict and control even if it is achieved by consent rather than coercion. The outcome of a conflict will depend on the relative strength or power of the contestants. Political conflict involves ideologies, institutions and interests. This can be illustrated by reference to conflict over environmental issues. There is *ideological conflict* between the environment and the economy represented in the conflict on the ground between conservation and development and the conflict over methods between the freedom of market forces and the intervention of planning. *Institutional conflict* occurs at different levels of authority between local and national government and international agencies. In every environmental conflict there will be opposing *interests* represented; on the one hand interests in profit, jobs and wealth creation and, on the other, interests in amenity, health, the conservation of natural resources and the survival of ecosystems.

These interests, though a source of conflict, are not always mutually exclusive. It does appear that the drive towards greater wealth and profit entails a deteriorating environment but it is not essentially so. Local environments are infinitely cleaner and healthier than they were in the times of gross pollution of the early industrial revolution. The threat to the global environment is, of course, much greater. Environmental interests are also economic interests. The conservation of resources and the survival and health of the population are essential for the maintenance of the economy. It is also possible for opposing ideologies to cohabit in the choice of policies for environmental management. It is possible to constrain the market with greater intervention in pricing and subsidies, as is proposed in the Pearce Report. It is necessary to understand the political conflict between ideologies and interests conducted at different institutional levels in order to explain the pattern of location of polluting activities and the choice of waste disposal pathways.

THE POLITICAL GEOGRAPHY OF POLLUTION

Some general trends in the politics of pollution may be observed as a background to the empirical discussion which follows. The most obvious is the increasingly successful resistance to locally unwanted land uses (LULUs) by local communities. This reflects the general increase in environmental concern and the consequent power of interests representing public health and amenity. It creates conflict between local and national levels of government.

When local protest is directed against the location rather than the project it is self-interested. It is an evocation of Not-In-My-Back Yard (NIMBY) feelings, implying that the same project located elsewhere would be acceptable. But, in some instances, this localized protest has also been linked to a more altruistic concern for public health and the conservation of resources or the survival of species. By focusing on specific issues it has been possible for local communities, acting in concert with more broadly-based environmental groups, to build powerful coalitions cutting across party, social and geographical barriers. These single-minded coalitions have achieved the power to defeat government proposals for waste management. When coalitions try to achieve a diversity of aims they are likely to fragment as conflicting loyalties or interests reveal their internal contradictions.

The success of environmental interests in mobilizing successful local resistance has restricted the opportunities for siting polluting and hazardous activities in greenfield locations. Tighter controls are imposing higher costs on industry which may seek locations in 'pollution havens' where controls are more relaxed. In these areas industry can expect a welcome for the jobs and wealth that it provides. The push from greenfield locations combined with the pull of 'pollution havens' reinforces existing locational patterns. Precise outcomes will vary depending on many variables which shift over time. The impact of political factors on location will be illustrated by a brief case study of the recent history of radioactive waste siting in the UK.

The increasing scale of pollution and the growing concern with environmental problems has added an international dimension to the political geography of waste and pollution. As we saw earlier, one consequence has been the increase of environmental inequalities between rich and poor countries. The rich countries have exported their unwanted polluting activities, as well as some toxic wastes (Chapter 12), to countries willing to accept them for the profit and wealth they bring. Thus, pollution havens are appearing all over the Third World. But the spread of pollution is also afflicting the wealthy nations who are fouling their own nest as well as polluting the nests of others.

These trends require urgent international political action. But, where the costs and benefits of pollution control are unevenly distributed, there are likely to be conflicts between nations. There is a recognized need to harmonize environmental standards. A second example, the conflicts over hazardous waste transfers, will be used to illustrate some of the political problems encountered.

THE POLITICAL GEOGRAPHY OF RADIOACTIVE WASTE

An unsolved problem

Political conflict over radioactive waste is far greater than for any other type of waste. Radioactive waste is created at all stages in the nuclear fuel cycle but particularly from nuclear reactors and reprocessing operations. There is no universal system of classifying radioactive wastes. Classification systems vary according to source, disposal route and levels of radioactivity. In the UK there are three broad categories; high-level heat-generating wastes (HLW) from spent fuel reprocessing; intermediate-level wastes (ILW) from processes closely related to energy production and fuel reprocessing including fuel cladding and control rods, sludges and resins (these are subdivided into long-lived intermediate wastes, mainly alpha emitters, with half-lives of over thirty years, and short-lived, mainly beta and gamma emitters, with half-lives under thirty years); low-level wastes (LLW) – the high volumes of lightly-contaminated wastes such as paper, clothing, laboratory equipment and the building debris of decommissioned facilities.

The fear of radiation is consistent over time and opinion polls indicate it is one of the public's greatest fears (Slovic *et al.* 1979a, 1979b). The burden of the nuclear waste is increasing and will persist for thousands of years. Although radioactive waste has been accumulating from nuclear reactors since the 1950s, it was not until 1976 that it was recognized as a problem urgently requiring a solution, when the Royal Commission on Environmental Pollution (RCEP) declared,

> There should be no commitment to a large programme of nuclear fission power until it has been demonstrated beyond reasonable doubt that a method exists to ensure the safe containment of long lived highly radioactive wastes for the indefinite future.
>
> (RCEP 1976, para 27).

Despite considerable effort on the part of government to deal with the problem of radioactive waste, virtually no progress in locating radioactive waste facilities has been made since the Report. The recent history has been one of successive retreats by government in the face of powerful opposition to plans for managing radioactive wastes.

Defeat on land and sea

In response to the RCEP's strictures, attention was initially focused on high-level wastes (HLW), primarily the spent fuel from reactors which was taken from power stations to Sellafield for reprocessing. An exploratory drilling programme was announced to assess the potential of sites in remote upland areas for a deep geological repository for the disposal of HLW. In every case the programme was thwarted by intense local opposition forcing the government to abandon the programme in 1981. Consequently, there are no immediate plans to find a suitable site for the eventual disposal of HLW. HLW waste in liquid form from reprocessing will be vitrified and remain at Sellafield to cool for about fifty years.

Two years later, in 1983, the government was forced to withdraw from dumping radioactive wastes at sea. The annual dump in the north-east Atlantic had provoked opposition from countries like Spain and Denmark and was the target of a well-publicized campaign by *Greenpeace.* This led to the transport unions, led by the National Union of Seamen, refusing to handle the wastes. This effectively ended sea dumping.

Shortly after this, in October 1983, the Secretary of State for the Environment, Patrick Jenkin, announced that on-land disposal of long-lived ILW was 'the safest and the best method provided that a site can be found with sufficient geological certainty and stability which will remain safe for the necessary period of time'. The site chosen was Imperial Chemical Industries' (ICI) disused anhydrite mine immediately underneath Billingham within the Teesside conurbation of half a million people, an area with a high concentration of hazardous industries. The proposal was opposed by all the local governments in the area and the protests were orchestrated by a pressure group Billingham Against Nuclear Dumping (BAND). Pressure was successfully applied on ICI, the owners of the mine, who declared that the proposal 'would not be in the company's best interests and is therefore opposed to it'. This proved decisive and, in January 1985, the government decided not to proceed with the Billingham project.

The power of protest

Within the space of four years, as a result of opposition, the government had perforce witnessed the abandonment of its policy for HLW disposal and for sea dumping. The options had been narrowed down to finding a method of on-land disposal of short-lived ILW and LLW. In October 1983, concurrent with the Billingham proposal, Elstow, in Bedfordshire, had been selected as a site 'most worthy of detailed investigation' for a shallow repository for short-lived ILW and LLW (NIREX 1983:1). This was needed since the only operating site for LLW, at Drigg near Sellafield, would be full before the end of the century. The project was justified on the basis of comparable experience elsewhere, notably in the USA. It was claimed that 'hydrogeological conditions are well-known' and that the site was coveniently situated both by reference to waste arisings and in terms of transportation facilities (ibid.:10). As in Billingham, there was united local opposition, led by the County Council and Bedfordshire Against Nuclear Dumping (BAND). They contradicted the NIREX claims on local geology, argued that experience elsewhere was not comparable and was highly flawed, and attacked the government for lacking any publicly-agreed strategy based on a comparative evaluation of alternative sites. The point about strategy was underlined by a House of Commons Committee which commented, 'the UK is still only feeling its way towards a coherent strategy' (House of Commons Select Committee on the Environment 1986:xii). The case for comparing sites was conceded by the Secretary of State in 1985, and in February 1986 he announced that three further sites, all in the clay lands of eastern England, at South Killingholme in Humberside, Fulbeck in Lincolnshire and Bradwell in Essex, in addition to Elstow, would be considered for a shallow repository.

Each community threatened with a repository established an action group. Opposition was co-ordinated by a national organization formed from the four communities plus Billingham called Britons Opposed to Nuclear Dumping (BOND). Three of the county councils involved (Essex excepted) formed a coalition to present a technical case against shallow burial and to lobby Parliament. In the aftermath of Chernobyl the government withdrew plans for the shallow burial of short-lived ILW, recognizing 'the gap between scientists' assessment of risks and the honestly-held perceptions of the local communities' (HMSO 1986: para. 32). But the government proceeded with plans to investigate the four sites by granting planning permission for exploratory drilling by means of a Special Development Order laid before Parliament.

Access to the sites was denied to the contractors by blockades mounted by protestors at each site. The drilling was delayed for two months while the protestors enjoyed considerable publicity for their cause. The County Councils Coalition toured nuclear waste facilities in Sweden, West Germany and France and concluded that those countries were 'far more advanced than the UK in the development of policies and practices for the disposal of radioactive waste' (County Councils Coalition 1987:1). A few months later, in May 1987, the Secretary of State, Nicholas Ridley, accepted the case put forward that co-disposal of LLW and ILW in a deep repository offered both a cost-effective and more publicly-acceptable solution. There was little doubt the immediate reason for the capitulation was the government's nervousness about the bad publicity it was receiving on the subject of radioactive waste on the eve of a General Election.

Retreat to the nuclear oases

Having lost all its options for radioactive waste management, the government had to begin all over again. Its policy had failed on two fundamental grounds – one, a lack of open debate to gain public credibility and, two, the lack of a strategy presenting the alternative options for consideration. Accordingly, in November 1987 the

Nuclear Industry Radioactive Waste Executive (NIREX) attempted to meet these criticisms in a consultation document called *The Way Forward.* It accepted a need 'for open discussion and feedback from a wide audience' and 'to promote public understanding of the issues involved' (NIREX 1987:4, 29). The document also established a wide area of search within which various deep disposal options on-land or beneath the sea-bed might be feasible. The consultation revealed very little support for a repository in any area though, initially, local councils in the areas surrounding Sellafield and Dounreay gave a cautious welcome to the approach. British Nuclear Fuels were investigating the geological potential for a site in the Sellafield area and NIREX were offered land for drilling by the Earl of Thurso in Caithness.

There was a political inevitability about the eventual selection of Sellafield and Dounreay as the two sites for initial investigation in March 1989. Nicholas Ridley confirmed this, saying, 'it would be best to explore first those sites where there is some measure of local support for civil nuclear facilities'. This support came from the dependent work-force in areas where the nuclear industry was the dominant employer. But there was also considerable local opposition in both communities. After a decade of futile attempts to secure locations for the disposal of nuclear waste, the government had retreated to its two main nuclear oases in the hope that there at least technical feasibility would coincide with political acceptability.

The political necessity of site selection

The recent history of radioactive waste in the UK illustrates the importance of political conflict in the siting of unwanted activities. The conflict was fundamentally between local communities defending their interests in amenity and health and avoiding the blighting effect of a nuclear dump. These interests cut across all parties and social classes. Ideological differences over nuclear power were submerged in the common interest of avoiding radioactive waste. It was a classic expression of NIMBY feelings. But the power of these local communities was increased by two features. One was that they struck a common chord with environmental groups operating on a national level whose opposition to nuclear energy was ideological. So long as the focus was on the specific proposals for nuclear waste the potential fragmentation of the coalition of interests could be averted.

The other feature was the tribulation suffered by the nuclear industry during this period. The problems of the Sellafield plant were given continual publicity. Moreover, bad publicity had an awkward habit of coinciding with radioactive waste initiatives. An accidental discharge of radioactive effluent into the Irish Sea and the concern about its link with local leukaemia clusters both coincided with the government's announcement of Billingham and Elstow in 1983. The catastrophe at Chernobyl occurred in April 1986 at a time when the battle between the four communities and NIREX was just being joined. The collapse of the nuclear energy option as a consequence of privatization in 1989 maintained the poor image of an industry trying to establish public confidence in its new approach to radioactive waste policy.

The nuclear industry and government had failed, first, to secure an acceptable solution for HLW, and, second, had abandoned sea dumping. They subsequently failed to achieve any of their on-land greenfield solutions. They are now left with a problem requiring solution. There is a national interest in the safe management of nuclear waste. It is in the nuclear industry's interest to find a solution if it is to regain public confidence that it can manage the waste it creates. There is apparently a consensus that the technical solution is deep disposal for ILW/LLW and, in the longer term, for HLW too. What is being sought is a similar consensus on the location. The strength of the communities in resisting sites, combined with the weakness of the industry in gaining public support, led ultimately to Sellafield and Dounreay.

The conflict over radioactive waste has increased the likelihood of the eventual selection of Dounreay or Sellafield. It is clear that resistance to greenfield sites is virtually unanimous whereas some support from dependent workers and their families can be guaranteed in nuclear oases. In addition, lessons have been learned and NIREX and the government are adopting the very principles of public debate and consideration of options advocated by their opponents. Above all, there is general agreement that the waste must go somewhere. Sellafield and Dounreay are politically isolated and people elsewhere will be relieved that sites have been identified. Sellafield and Dounreay are the hapless victims of political conflict. Only insuperable technical problems are likely to rescue one or the other from being the repositories of the country's intermediate- and low-level waste.

In the case of radioactive waste the importance of the relationships between politics and place are clear. Each community is defending its unique qualities but making alliances with others where this is likely to prove fruitful. Provided the technical options offer a wide range of potential locations, the eventual choice will be essentially political. This same principle operates between countries as the next case study demonstrates.

THE POLITICAL GEOGRAPHY OF WASTE

A neglected issue

Radioactive waste apart, waste management has failed to arouse great political interest. In consequence there has been a lack of public information and a general neglect of the issue manifested in a lack of planning, the existence of 'cowboy' operators, and inadequate inspection and monitoring. There has been a lack of information on waste arisings and the composition of waste streams, a point underlined by Parfitt's analysis in Chapter 3. In its opening remarks in a Report on Toxic Waste, the House of Commons Environment Commitee conveyed a sense of frustration and alarm at the situation it had encountered,

> Never, in any of our enquiries into environmental problems, have we experienced such consistent and universal criticism of existing legislation and of central and local government as we have during the course of this inquiry.
>
> (House of Commons Select Committee on the Environment 1989:xi)

The report noted there were only five inspectors for over 5,000 disposal sites. It described as 'scandalous' the fact that only twenty-three out of seventy-nine waste disposal authorities in England had deposited plans by 1989, over a decade since the requirement to do so was made. Poor management practices at different sites had been revealed in successive annual reports by the Hazardous Waste Inspectorate. Despite some improvement, the Third Report in 1988 concluded that 'the contrast between the best and the worst remains as dramatic as ever' (Hazardous Waste Inspectorate 1988:83).

There is evidence that public concern is increasing. The lurking dangers of toxic substances leaking from abandoned or poorly maintained tips were emphasized in an unpublished government survey in 1974 which was referred to in a government circular in 1975 and publicized in a report by Friends of the Earth which appeared in the

Observer in February 1990. The report identified 1,300 sites (a quarter of the total toxic tips) from which toxic materials could be leaking into groundwater and fifty-nine from which there was a 'serious risk' of contamination.[4] It is an interesting comment on the rise in public interest that a report which passed without much comment in 1975 should achieve such fame fifteen years later. Indeed, public concern about toxic waste has focused on various incidents in recent years. As we saw in the previous chapter, the vessels *Karin B* and *Deep Sea Carrier*, carrying toxic wastes, were refused entry to British ports; vessels carrying polychlorinated biphenyls (PCBs) from Canada were also turned away; and proposed shipments of domestic refuse from the United States to Cheshire were also forestalled. In each case the trade was perfectly legal but the force of protest was sufficiently great to encourage the authorities to ensure that procedures were devised to prevent the imports.

So far the UK has not experienced the kind of transforming incident that provokes sustained public alarm and attention and ensures government action and money to prevent recurrence. In the United States such an event was the evacuation of Love Canal in 1978 which led to the establishment of Superfund to ensure the clean-up of contaminated sites (see Chapter 12). In the Netherlands the turning-point was in 1980 when houses at Lekkerkerk, built on a reclaimed waste site, had to be abandoned as a result of illnesses caused through contamination by illegally dumped toxic wastes at the site. This led to the Dutch approach of *multifunctionality* requiring land to be cleared up so that any land use is possible. Commenting on the UK's experience the House of Commons Environment Committee had this to say,

> It is undoubtedly the case that the UK has been spared some of the worst effects of uncontrolled dumping – though our recent inquiry on the disposal of toxic waste has not led us to be complacent about the quality of waste management in this country.
>
> (House of Commons Select Committee on the Environment 1990:xx)

The impression of unevenness in the quality of waste management and the need for better information and control is supported by the findings in this book. There is an emphasis on improved management information (Chapters 3, 6), the need to focus on environmental benefits (Chapters 1, 2), and the need to be aware of the hidden consequences of decisions (Chapter 5).

A change of approach

The government responded to this general climate of concern and criticism in the publication of its Environmental Protection Bill in 1989, enacted in 1990. The Act introduced major changes in the principles of waste management and pollution control. The concept of Integrated Pollution Control (IPC) had already been embraced with the merger of the hitherto separate Industrial Air Pollution Inspectorate, Controlled Waste Inspectorate and the Radiochemical Inspectorate into Her Majesty's Inspectorate of Pollution (HMIP) in 1987, later joind by the new water inspectorate established when the water industry was privatized. The traditional pragmatic British system of applying the 'best practicable means' (bpm) to achieve 'presumptive limits' is replaced by the principle of 'Best Available Technology Not Entailing Excessive Costs' (BATNEEC) which places the emphasis on the application of controls rather than achieving an appropriate balance between control and cost which was implicit in bpm. Instead of retrospective prosecution or persuasion, companies will be given 'prior authorization' to operate within legally-binding emission limits. There are strong powers of enforcement and a residual duty to apply controls to processes not otherwise covered.

Changes in the role of waste disposal authorities are intended to tighten up the licensing and enforcement of waste management. The 'duty of care' has also been introduced obliging all holders of waste 'at every stage in its history' to take all reasonable measures 'for ensuring that controlled waste is not illegally managed, that it does not escape from control, that it is transferred only to an authorised person and that it is adequately described to enable proper handling and treatment' (Department of the Environment 1990: paras 6 and 7).

The change of approach to pollution control brings the British system more into line with that adopted in other EEC countries. The British system was criticized for encouraging a cosy relationship between the inspectorates and industry. But it could also be defended as a flexible system able to apply principles of equity and efficiency which respect the needs of industry while protecting the environment (Hawkins 1984; Vogel 1986). The new system is more open and specific and focuses on common standards of pollution control and environmental quality objectives. A major problem will be controlling smaller companies who will find it difficult to comply with the standards required of a more coercive regime. In any event there is the danger that the system will fail to achieve its objectives because it is under-resourced and over-centralized. The resources of HMIP are unlikely to be able to cope with the increased demands. Meanwhile, local authorities, starved of resources and with limited powers, will be unable to respond effectively to the increasing environmental concerns of their electorates.

Locational outcomes

Although toxic wastes have not yet created as much public concern as radioactive wastes, there are similarities in the locational outcomes. Local anxiety has been aroused by specific pollution incidents or the fear of leakage or accidents from toxic tips or provoked by proposals for new disposal sites. Local action groups have sought clean-up and compensation for specific problems such as Love Canal. In the UK local pressure has resulted in tougher conditions being imposed on waste disposal operations. There have been instances where opposition has succeeded in preventing the importation of toxic wastes into landfills. At a more regional or national level there is concern about the safety of disposal sites and the dangers from transportation of toxic materials.

It is the transboundary pollution problems caused by the discharge and transportation of toxic wastes that has secured the most sustained political attention. Environmental groups, notably *Greenpeace*, have drawn attention to the trade in toxic wastes. The European Commission has been particularly active in developing a series of Directives designed to introduce a comprehensive regulatory framework for the management of toxic wastes and pollution binding on all member countries.

The political pressures are drawing the net more tightly around the polluters narrowing the spatial options for disposal of toxic wastes. Landfill space is becoming scarce and incineration is expensive and restricted to a few facilities. As regulation tightens so the outlets for toxic wastes are reducing. The internal free market in the EEC may discourage national governments from developing new disposal facilities for fear of being swamped with waste from other member countries. The lack of low-cost options will lead many waste disposal companies, especially the more unscrupulous, to seek outlets further afield. Among the most obvious locations will be those in Third World countries. But the political changes in eastern Europe may open up the possibilities of dumping toxic wastes in countries where pollution is already high and controls are lax. Although the locational options within western Europe and North America may be narrowing, new ones may be opening elsewhere.

SUSTAINABLE OPTIONS

The notion of 'sustainability', ensuring that the present use of resources does not rob the future, has become politically fashionable. There is a growing consciousness that the pressures imposed on the resources of the environment were becoming intolerable, that the limits of sustainability were being reached.

The intuitive response of politicians has been rhetorical, a signal that they have understood the political significance of environmental issues. It has, of course, been more than that. After all, the 'environment' comprises a set of issues which have long required the routine attention of governments. Policies for dealing with such issues as waste disposal or air pollution are a necessary part of a government's responsibility to protect its citizens and the natural environment. What is relatively new is the widespread anxiety about the long-term regional and global environmental implications of contemporary industrial and agricultural processes. The damage to health and the environment that can be inflicted by radioactivity, by toxic wastes and by the pollution derived from wastes, can threaten sustainability at local, regional and even global levels. These long-term problems are in direct conflict with the political imperative of maintaining growth and prosperity in the short term. The sustainability of the environment is incompatible with the sustainability of contemporary processes of economic development and growth.

Sustainability requires that wastes and pollution are so managed that they do not cause a threat to the quality of the environment. In this book we have considered the location of waste production, transportation and disposal. The geographical perspective, with its emphasis on spatial analysis and locational constraints and opportunities, provides a synthesis of the factors involved in making locational decisions but, geography has been lamentably weak in its understanding of the political factors which have narrowed the locational options available to companies and governments.

The political options for dealing with environmental problems such as waste and pollution are narrowed in two ways. First is the political imperative of dealing with the short term. Governments depend on industry for wealth creation and taxation and for satisfying consumer demands. Democracies (and, in less evident ways, other political systems too) depend for their survival on their ability to deliver economic prosperity and growth. Governments fully understand that environmental controls must prove to be politically feasible (Lindblom 1977). They cannot, therefore, impose controls which will be ignored or evaded or resisted by industry.

A second limitation on the available options is caused by the pressure from the public for protection from environmental risk. As we have seen, local resistance to unwanted land uses has achieved considerable power and can be greatly strengthened by the development of broad coalitions reflecting more general concerns about the environment. Governments try to find the balance which ensures that the immediate demand for economic prosperity can be met without offending the long-term requirement of environmental protection.

In practice governments have achieved this balance by taking the political soft options. In the case of waste location there has been a retreat to the pollution havens and nuclear oases. These are areas familiar with the polluting industry, where the economic wealth and jobs created would be welcomed. As greenfield sites are resisted, so the tendency is towards increasing environmental inequalities. This process is now occurring on a global scale. In the developed countries waste and pollution management is becoming more costly, more regulated and the options are reducing. Waste producers are seeking outlets elsewhere and dumping difficult wastes in some parts of the Third World. Even here, as we saw in Chapter 12, the options are narrowing as countries react collectively to this evidence of environmental exploitation.

It is no longer simply a matter of locating waste and polluting activities so that their harmful effects are confined to specific areas. As the scale of technology increases, it is becoming difficult to avoid the impacts of the wastes dispersed into the soil, atmosphere, rivers and oceans. Furthermore, wastes are emitted in liquid or gaseous form from a myriad of stationary sources but also from mobile sources in the form of motor vehicles. Waste residues can travel far from their source polluting pristine environments. In some cases waste residues may threaten large areas with environmental devastation or pollute the atmosphere with global implications. The indiscriminate impact of transboundary pollution is an international political problem of immense complexity since it is impossible to determine precise causes and effects and costs and benefits. The likelihood of denying short-term national economic self-interests in order to avoid a possible long-term environmental catastrophe is extremely remote. But, as Chapter 10 warns, the cost of underestimating the possibility of a catastrophe may be so great that it is best to assume it will occur and to take the necessary avoiding action. The issue becomes one of not where we locate waste-creating and polluting activities, but whether we can reduce or eliminate certain processes altogether. By the time the burden of waste and pollution threatens environmental degradation and human health on a global scale, the options will have narrowed to the point where there is no choice.

NOTES

1 Environmental Data Services Ltd (ENDS), Report no. 176, September 1989, p. 3. The main environmental issues were chemical pollution of rivers and seas (64 per cent), sewage pollution of bathing waters (59 per cent), radioactive waste (58 per cent), ozone depletion (56 per cent), oil spills (53 per cent).
2 Details in *USA Today*, 30 November 1989. Atmospheric pollution (smog, ozone depletion, global warming, acid rain) ranked highest among environmental concerns, followed by water quality, then waste disposal.
3 Reprinted in *Science and Public Affairs*, 1989 (4), 6.
4 The *Observer*, 'Britain's Buried Poison', Colour Supplement, 4 February 1989. The original 1974 study was released by the Department of the Environment in 1975 and was referred to in Circular 39/76, *The Balancing of Interests between Water Protection and Waste Disposal*, 13 April 1976.

REFERENCES

County Councils Coalition (1987) *The Disposal of Radioactive Waste in Sweden, West Germany and France*, prepared for the County Councils Coalition by Environmental Resources Ltd, January.
Department of the Environment (1990) *Waste Management: the Duty of Care*, A consultative paper and draft Code of Practice under the Environmental Protection Bill, February.
Hawkins, K. (1984) *Environment and Enforcement: Regulation and the Social Definition of Pollution*, Oxford: Clarendon Press.
Hazardous Waste Inspectorate (1988) *Third Report*, London: HMSO.
HMSO (1986) *Radioactive Waste*, The Government's Response to the Environment Committee's Report, Cmnd 9852.
House of Commons (1986) Environment Committee, First Report, Session 1985–6, *Radioactive Waste*, London: HMSO.

House of Commons (1989), Environment Committee, Second Report, Session 1988–9, *Toxic Waste*, London: HMSO.
House of Commons (1990) Environment Committee, First Report, Session 1989–90, *Contaminated Land*, London: HMSO.
Lindblom, C.E. (1977) *Politics and Markets: the World's Political-Economic Systems*, New York: Basic Books.
Nuclear Industry Radioactive Waste Executive (NIREX) (1983) 'The disposal of low and intermediate-level radioactive wastes: the Elstow storage depot', a preliminary project statement, October, Harwell.
Nuclear Industry Radioactive Waste Executive (NIREX) (1987) *The Way Forward*, a discussion document, November, Harwell.
Pearce, D., Markandya, A. and Barbier, E. (1989) *Blueprint for a Green Economy*, London: Earthscan Publications.
Royal Commission on Environmental Pollution (RCEP) *Sixth Report, Nuclear Power and the Environment, The Flowers Report*, Cmnd 6618, London: HMSO.
Slovic, P., Fischoff, B. and Lichtenstein's, S. (1979a) 'Rating the risks', *Environment* 21, April, 114–20.
Slovic, P., Fischoff, B. and Lichtenstein, S. (1979b) 'Images of disaster: perception and acceptance of risks from nuclear power', in G.T. Goodman and W.D. Rowe (eds) *Energy Risk Management*, London: Academic Press, 223–45.
Vogel, D. (1986) *National Styles of Regulation: Environmental Policy in Great Britain and the United States*, Ithaca and London: Cornell University Press.
World Commission on Environment and Development (1987) *Our Common Future*, Oxford: Oxford University Press.

Index